W9-ARH-028

Jacques Cousteau

JACQUES
COUSTEAU

THE SEA KING

Brad Matsen

Pantheon Books, New York

Library of Congress Cataloging-in-Publication Data

Matsen, Bradford
Jacques Cousteau : the sea king / Brad Matsen.
p.   cm.
Includes bibliographical references and index.
ISBN 978-0-375-42413-7
1. Cousteau, Jacques Yves.   2. Oceanographers—
France—Biography.   I. Title.
GC30.C68M38   2009   551.46092—DC22   [B]   2009011640

www.pantheonbooks.com

Printed in the United States of America

First Edition

2   4   6   8   9   7   5   3   1

FOR LAARA

I believe in children. I live for children.

Jacques Cousteau

CONTENTS

ILLUSTRATIONS

PREFACE

A few weeks after I decided to write about Jacques Cousteau, I went to St.-André-de-Cubzac, France, where he was born and is buried. It's a typical market village on the fringe of Bordeaux with a busy highway running through the center and narrow streets off to each side that are deserted during the day because almost everyone works in the city. The second-floor bedroom in which Cousteau first drew breath is now part of an apartment over a pharmacy, across the street from the windowless stone abbey that anchors St.-André to the twelfth century and the Roman Catholic church. Two blocks west, I found a civic monument to Cousteau at the traffic circle on the approach from Bordeaux. Above a splash of carefully tended annuals in summer bloom, a twice-life-size wooden dolphin is mounted 10 feet off the ground on a steel pole. Rendered by the sculptor in the act of leaping from the sea, the dolphin holds in its mouth a red knit watch cap like the ones worn so famously by Cousteau and the crew of *Calypso*. On a separate pole, arrowed signs guide visitors to the local highlights—the Philippe Cousteau Professional Academy, City Hall, and the Forty-fifth Parallel Ecological Observatory, one of several founded in the eighteenth century to study the earth precisely halfway between the equator and the North Pole.

Just past the traffic circle is the cemetery, two acres surrounded by limestone walls beyond which the ancient vineyards roll like ocean swells toward the Gironde Estuary and the Atlantic. On the late August afternoon I was there, the walls were so bright with reflected sun that it was hard to look directly at them. Inside, there were rows of tombs, some of them like blockhouses intended to protect their occupants, some like ornate temples decorated with stone flowers, angels, and portraits of the deceased. A sign at the gate pointed the way to *Sépulture Cdt. Cousteau*.

Calypso, *August 2005, La Rochelle, France* (COURTESY OF THE AUTHOR)

Cousteau's parents, Daniel and Elizabeth, lay together in a knee-high chamber of limestone bricks capped by a simple cross inscribed with a single word: COUSTEAU. A carpet of red flowers covered the center of the tomb, around which were conical, waist-high evergreens spaced like sentries. To the side, facing over the vineyards just then plump with the season's crop, were three graves marked by wrought iron crosses. Over one of them was a foot-square piece of slate with weathered gold leaf lettering:

J Y Cousteau
Papa du Globe

I was alone at the tomb for only a few minutes before a man and woman I guessed to be in their midthirties arrived. They looked at me but didn't speak, so I moved away to give them the privacy they seemed to want. While the man stood silently with his hands clasped in front of him, the woman reached out and swept a few dry leaves from the top of the tomb. Then she picked up a handful of pebbles from the path at her feet and arranged them in the shape of a heart on the spot she had cleared. Around her heart were similar commemorations of pebbles—an anchor, a ship's hull, another heart, the letters *JYC*, a circle. When the woman turned to face her husband, I saw that her eyes were shiny with the beginnings of tears. She shrugged as though slightly embarrassed by her emotions and took his arm. They nodded at me as they started for the cemetery gate. "Did you know him?" I blurted in English. "No," the woman said. "But we loved him."

In the 1950s, '60s, and '70s—the middle decades of my life—Cousteau was the most internationally recognizable television star on earth. His success as a filmmaker had peaked with *The Undersea World of Jacques Cousteau,* which aired four times a year for a decade before being canceled by ABC in 1976. After that, Cousteau never again reached television audiences of tens of millions, though he produced over a hundred more documentaries. Instead, he became a revered elder of the environmental movement, credited with kindling a new awareness of the need for stewardship of the world's oceans and rivers. The man and woman in the cemetery were thirty years younger than I, members of the generation after mine. The woman arranging her pebbles on his tomb added a layer of complexity to a man I regarded as a brilliant showman who had coinvented the Aqua-Lung because he wanted to breathe underwater, had made movies to show the world what he found, and had used his celebrity to transform the human relationship with our planet. I had not understood that Cousteau was, simply and timelessly, beloved.

The next day, I took an afternoon train north along the Atlantic coast to the port city of La Rochelle to look for what was left of Cousteau's *Calypso,* the converted World War II minesweeper that had become the most famous research ship in history. At the dock of the Maritime Museum, I found a wreck that bore only a vague resemblance to the heroic white ship I remembered from many hours sitting in front of the television eating a cooling turkey TV dinner and watching Cousteau and his divers explore the underwater world.

The hulk was weeping rust streaks from corroded fittings on its sooty white flanks. Frayed dock lines seemed to be straining to keep it afloat. The name and hailing port had been covered over with off-shade paint that looked as if it had been applied in a rush to conceal its identity. The transom planks, darkened by rot, bulged ominously. In four places, flat canvas slings were wrapped tightly around the hull as though to contain its innards. On deck, a clutter of metal tubing, wire, gas canisters, and the remains of a crow's nest in front of the pilothouse looked dangerously jagged and forlorn. The boarding gate was missing, leaving a hole in the gunwale through which I could see the ship's docking ladder under a pile of crumpled sheet metal. On the side of the ladder I made out sun-bleached green letters in uneven Greek script with the letter *a* in the word *Calypso* replaced by the symbol of a fish: α.

La Rochelle has been in continual use as a port since the tenth century, when it dealt mainly in red wine and salt. From the sea, ships have to navigate channels through treacherous, shifting sand flats at the confluent mouths of the Dordogne and Garonne rivers, but once inside the harbor, they are safe from all but the worst storms. As I looked at the wreck at the dock, I realized that this perfect harbor was very likely the last port of call for a ship known to more people than Jason's *Argo,* Jules Verne's *Nautilus,* Captain Cook's *Endeavour,* or Ernest Shackleton's *Endurance.*

Cousteau in his tomb in St.-André had seemed natural to me. We all owe a death. Boats eventually die, too, but the wreck of *Calypso* felt wrong, neglected, and dishonored. I had already picked up faint rumblings about a legal stalemate over possession of the ship and bitter conflict among Cousteau's survivors. The sight of *Calypso* rotting at the dock raised questions I knew I would have to answer to represent his life accurately. How could a man of such immense power have allowed his former mistress and widow, his surviving son, and their children to descend into bitter conflict over his legacy? Was he a tragic character hidden behind the veil of celebrity? Does he deserve our enduring love?

At dusk, after a long hour absorbing what the wreck of *Calypso* was telling me, I walked over to the Maritime Museum and asked one of the ticket takers what was going to happen to Cousteau's famous ship. I hoped that its presence at the museum dock meant that there were plans to rebuild it as an exhibit. Shaking her head as though she was angry to speak the words, the ticket taker said, "She will not be repaired. She is destroyed."

For three years after my visits to St.-André and La Rochelle,

Jacques Cousteau dominated my end of conversations, an indulgence that is familiar to anyone who spends much time with a writer. I talked often about the scene I witnessed at his grave and my sadness and consternation about the wreck of *Calypso*. Without exception, I received in return stories of memorable moments in the lives of friends, colleagues, and strangers in which Cousteau and his adventures played a role. Everyone, of course, could do the accent. Some told me that he had changed their lives forever.

The owner of a wineshop around the corner from my hotel in Bordeaux—a woman who wore her five or six decades on earth like a designer dress—told me that as a teenager she had seen Cousteau on television and decided she would marry him. When she found out he was already married, she picked a man who looked enough like him to be his brother.

I mentioned Cousteau to a docent at the Picasso Museum in Antibes, who told me that the first time Cousteau met Picasso, he gave the artist a piece of polished black coral from the Red Sea. Picasso had it in his hand when he died.

It was like telling people about your operation or a recent death in the family. Everybody had a story. A marine biologist from Olympia, Washington, said he had watched every episode of *The Undersea World,* decided to devote his life to studying the sea and its creatures, and never wavered in his passion. He gave me a program he had saved from a Cousteau Society environmental rally held at a sports arena in Seattle in 1977. An oceanographer told me about going to the same rally in Seattle, where the image of Cousteau walking across the basketball court was indelibly etched in her memory. She said he was like a powerful, kind, and handsome king. Another woman remembered that her father, who was a diver and sea captain, insisted that his wife and children watch Cousteau on television. He wanted them to understand why he was so excited about what he did for a living. A champion swimmer from California told me that after he watched Cousteau specials as a teenager, he majored in biology in college, became a master diver, made more than fifty documentaries about the ocean, and helped found one of the world's great aquariums in Monterey. A thriving novelist with a Ph.D. in marine science told me that his affection for Cousteau as a boy had sent him to sea to be an oceanographer, but it had proved to be the wrong life for him. One thing nobody told him, he said, was what hard work research diving really is. Watching Cousteau, he had not been able to imagine it as anything but fun.

Since I began work on this book, I have heard hundreds of these testaments. My own includes seeing Cousteau's *The Silent World* at an army base movie theater and, a few summers later, buying an Aqua-Lung in partnership with a friend so we could spear black sea bass to sell to neighbors in our Connecticut seaside town. After that, I spent most of the next forty-five years writing about the sea and its creatures. In the 1980s, partly because of Cousteau's booming insistence that human beings were doing irreparable harm to the oceans by over-fishing, I guided the editorial policy of the country's most respected commercial fishing magazine in the direction of environmental responsibility. The concept that without fish there could be no fisher-men finally started to make sense to fleets that had, until then, fished with little restraint, driving some species into total collapse.

Not everyone I met in the course of my research was kind to the memory of Jacques Cousteau. From his earliest encounters with his collaborators in the invention of the Aqua-Lung, he possessed the power to persuade other people to join him in fulfilling his vision. En-gineers, sailors, divers, directors, producers, writers, reporters, friends, family, and lovers willingly helped him realize his visions and were grateful for the chance to share even a sliver of his adventures. Later, many of those people felt slighted as they disappeared in his wake while he became famous and moved on. Most of them were grateful to have been part of Cousteau's magnificent journey, but few believed that they had ever known the man at all.

As Peter Guralnick, the author of a brilliant biography of Elvis Presley, wrote about trying to understand the life of a celebrity: "No matter how long one peers in from the outside, it is never quite the view from within." Instead, guided by the hundreds of books, maga-zine articles, newspaper stories, and films that constitute Cousteau's oeuvre, I have constructed a timeline for his life and a chronicle of his inventions, adventures, and achievements. In interviews with friends, former *Calypso* crew members, and some of his family, I tried to go beyond that public man to understand his nature and motivations, perhaps even to know him.

This book is a voyage into the life of Jacques Cousteau, necessarily burdened by the flaws contained in the opinions of dozens of people who knew him or wrote about him. It is a biography of sorts, but not a day-by-day recitation of the events of his life. Nor is it the simplistic account of the life of a man who was far more complex than most of us, whose true character was hidden behind the veil of celebrity. I offer to all those whose lives were changed by Jacques Cousteau a few things they might not know about him. Most of all, I intend this book to be a respectful, honest remembrance of the man who brought the oceans and rivers of the world to life for all of us.

JACQUES
COUSTEAU

PRELUDE

Autumn 1977

The road to paradise is paradise.
—*Jacques Cousteau, citing a Spanish proverb*

THE SOUL OF Jacques Cousteau's code for living was his belief that reflecting on the past is useless. Once, when one of his sons came to him for help making a biographical film to celebrate his birthday, Cousteau said, "You're wasting your time talking about my past. Don't count me in. I can't help you."

In the autumn of his sixty-seventh year, therefore, Cousteau would not have agreed—or cared—that what can now be understood to be the enchanted opening movements in the symphony of his life were ending and a final dissonant movement was about to begin. He had just finished a two-month fund-raising tour of the United States, where sellout sports arena crowds in six cities had welcomed him like a visiting monarch and opened their checkbooks for the television celebrity.

Despite Cousteau's insistence in a recent interview that he did not feel responsible for anything, he was awash in responsibilities. He was the president of the world's largest manufacturer of scuba diving equipment, the director of the Musée Océanographique de Monaco (Oceanographic Museum of Monaco), the director of the French Office of Undersea Technology, the founder and president of the fastest-growing environmental organization in history, and the president of a television production company that between 1966 and 1976 had made him one of the most recognizable people on earth as the star of *The Undersea World of Jacques Cousteau*.

His days were as heavily scheduled as those of a head of state, leaving little time for anxiety, but that fall Cousteau was very worried

Cousteau monument, St.-André-de-Cubzac, France

about his future in television. The American Broadcasting Company had canceled *The Undersea World* after a ten-year run, even though it was among the most successful television ventures of all time. ABC and the other two commercial networks, NBC and CBS, had simultaneously decided that science and nature documentaries, variety revues, and quiz shows weren't drawing big enough audiences. They were turning instead to an addictive new formula built around weekly serial dramas called situation comedies.

Within weeks of the cancellation, Cousteau had proposed a new series to the fourth American television network, the Public Broadcasting System, a loose new conglomerate of viewer-supported stations across the United States. After endless presentations to corporate boards of directors, he had finally found a sponsor, the Atlantic Richfield Petroleum Company (ARCO). PBS audiences were minuscule compared with those of the major networks, but he knew he had no place else to go on television. Everything for which he had worked as a filmmaker and explorer for a half century was lost without it.

Radio and television had fundamentally altered exploration, allowing listeners and viewers to share moments of discovery instantaneously with heroic men and women in faraway places. The live broadcast of exploration had begun in August 1932, when oceanographers Otis Barton and William Beebe had themselves sealed in a 4.5-foot steel ball and lowered three-quarters of a mile into the Atlantic to peer through 6-inch portholes into the inky darkness. They reported a full hour of their dangerous journey as it was happening on the new National Broadcasting Company radio network. More than fourteen million people had been riveted to their radios in England and the United States. Vicarious exploration had reached its zenith on July 20, 1969, when one-eighth of the three and a half billion human beings on earth witnessed Neil Armstrong and Buzz Aldrin walk on the moon.

Like them, Cousteau had been catapulted to stardom by the astonishing new medium that sent information around the world on invisible electromagnetic waves. Now, for the first time, he was faced with adapting to its vicissitudes. Incredibly, lunar astronauts had also experienced the fickleness of television networks and audiences. Every person on earth with access to a television set had watched Armstrong and Aldrin walk on the moon. Very few of them could name the astronauts who made the last moon landing just three and a half years later (Gene Cernan, Ron Evans, and Harrison Schmitt).

His new television series, *The Jacques Cousteau Odyssey,* consisted of a dozen episodes built around underwater archaeology, shipwrecks, and environmental disaster airing over three broadcasting seasons. The first show, on the discovery of the sunken ocean liner *Britannic,* drew favorable reviews and a solid audience, but the costs for filming the series were outrunning ARCO's money. Cousteau needed more than $2 million a year to keep his two hundred employees, companies, expeditions, and institutes afloat. He didn't particularly care about money as long as he had enough, and his chief financial tactic was simply going out and getting more cash when he ran out.

The six-week fund-raising tour in the United States organized by the Cousteau Society, a nonprofit corporation he had launched four years earlier, would pay only some of the bills. In six cities, he and his son Philippe gave speeches to sellout crowds in sports arenas. At the finale in Seattle, more people packed a basketball coliseum than had come to the same place to see the Rolling Stones a few weeks earlier. Before his lecture, as he had in each city, Cousteau set aside time to listen and talk to schoolchildren. They gave him flowers and drawings of seals, dolphins, whales, and one in which *Calypso* was depicted with wings hovering over the sea surrounded by a silver aura. After the gifts, Cousteau told the children they could ask him anything they wanted to about his life.

How old are you? Sixty-seven. How deep is the deepest you ever dived? Three hundred and seventy-two feet. Have you ever swum with a blue whale? I'm sorry to disappoint you, but I've never swum with a blue whale, because there are very few of them left anymore. What is it like underwater? It's fantastic underwater because it is like floating in space. Are whales smarter than humans? Some whales have bigger brains than humans, but that doesn't mean they are smarter. A girl raised her hand. Cousteau nodded to her. Without the slightest hint of humor, she said, "When I grow up, I want to marry a whale." Another child told him that his school was changing its name to Orca Elementary to honor Cousteau and his work saving the whales. Cousteau beamed. "Wonderful," he said. "That's beautiful."

After the children, Cousteau sat down with two reporters from Seattle's antiestablishment newspaper.

"What does your work—the books, films, and the Cousteau Society—mean to you personally, as Jacques Cousteau?" one of them asked.

"This is an introvert question and I am not introverted," Cousteau replied. "I am extroverted. I do not find my pleasure in asking questions about myself. I find my satisfaction in dealing with questions that concern the community and the outside world. We are more and more induced by publicity and the media to turn toward our neighbor. I hate my neighbor." (He laughed.) "I like to look to the outside world."

"You must have a sense of responsibility," the other reporter said.

"I hate responsibility," Cousteau snapped. "I feel in gear with the life of the world and that is not the same thing. The sense of responsibility is introverted, it gives you an importance. None of us has any importance, but rather we are in a symphony. The man who plays the violin in the symphony, he does not have a sense of responsibility. He is cooperating . . . Life is a symphony and we are playing a tune in the symphony; there is no responsibility there."

"Do you believe in destiny?"

"No."

"What about God?"

"If there is anything like God, it is so complex that we have no idea of what it is like. The concept of God is separate from ourselves. We have no importance to a God if there is one."

"What do you rely on?"

"Nothing."

"Don't you have some sense of faith?"

"No. I believe in the *instant*. I am going to give you a quote that has guided my life. I don't like quotes, but this one enlightens me. It's a Spanish proverb: 'The road to paradise is paradise.' "

Cousteau savored his celebrity and the freedom to roam the planet, but remaining on television was crucial because it was the most powerful medium on earth for sounding an alarm. While he was in the United States, *Calypso* and her crew were at sea, sampling the water and bottom sediment for pollution by heavy metals and other toxins that might account for the dramatic collapse of life in the shallows near shore. Not long ago, the Mediterranean basin had been the entire world to the civilizations that bloomed on its shores, a gift that had nourished the Samarians, Persians, Greeks, Egyptians, Gauls, Romans, and the countless other tribes and bands that had been fed and cleansed by its waters. Now, even the most casual observer could see that the nearly landlocked ocean known to the Romans as Mare Nostrum was headed for disaster. The lush habitat for plants and animals was becoming a polluted soup spoiled by the refuse of tens of millions of people crowding its shores. Cousteau had become obsessed with sounding the alarm about the Mediterranean and the rest of the world's oceans, which, thanks to him, had been revealed as fragile beyond anyone's wildest imagination. As exploring and filming the underwater world had dominated the first half of his life, saving what he had seen would command his passion for the rest of it.

"The Mediterranean will be the first to die," he told a French magazine reporter, "and become a warning for the world."

Cousteau was the secretary general of the International Commission for Scientific Exploration of the Mediterranean, chairman of Eurocean, a joint venture of twenty-four European companies to explore and preserve the oceans, and director of the Oceanographic Museum of Monaco. He had easily convinced all three to support the survey expedition, accomplishing several goals at once. The voyage to measure pollution would become an episode for the PBS series in which film footage of gorgeous coral and swarms of fish that he shot forty years earlier were contrasted with images of his divers descending into today's bleak underwater wasteland. It would help make the recovery of the Mediterranean Sea a cause célèbre, and promote the work of the Cousteau Society. Because of television and Jacques Cousteau, millions of people would know the results of the Mediterranean pollution survey that just a decade earlier would have been shared only by a handful of scientists. Cousteau was outraged by the dismal state of the ocean of his youth. He had a dark sense of foreboding about the overall health of a planet that was supporting five times as many people as it had when he had been born in 1910. Increasingly, he was desperate to transmit that message to the world.

Cousteau's own children and grandchildren would inherit the misery of unimaginable privation and sadness if his dire predictions about the earth came true. Both of his sons and their families were living in Los Angeles, where, for a decade, the day-to-day work of producing and editing his television show made it as much a home as any of them ever had. His younger son and coproducer of the *Odyssey* series, Philippe, was at that time recovering from a broken leg suffered in the crash of his gyrocopter while filming on Easter Island. His other son, Jean-Michel, was an architect working on aquarium exhibits and lecturing. Both his sons were married, with children of their own, and Cousteau believed that their prospects for the future were grim unless humanity ceased to be the plague on the earth that it had become.

In the autumn of 1977, while Cousteau was absorbed in the passions of his sacred present, events in the unknowable future were about to change his life forever. He could not have known, for instance, that he was about to become the patriarch of a second family. His affair with an Air France stewardess named Francine Triplet seemed to be more than just another of his endless liaisons with women. Within a year, they would have a child together, then another, beginning a secret life that would remain hidden for almost fifteen years. Cousteau also could not have known that soon he and his wife, Simone, would suffer their ultimate agony together, the death of a child. Afterward, Cousteau carried on while Simone retreated alone to *Calypso,* where she was known as *La Bergère,* The Shepherdess. Later, someone asked Cousteau if it had been difficult for him to be the commander of *Calypso* during her halcyon days.

"Not if Simone was on board. She was the cook, the mother of thirty sailors, the one that advised, the one that ended the fights, the one that told us when to shave, the one that challenged us to do our best, the one that we counted on, our best critic, our first admirer, the one who saved the ship in a storm. She was the smile each morning and the warm good night. *Calypso* could have lived without me, but not without Simone."

1

LA BERGÈRE

Marriage is absolutely archaic. It is a device people
use to avoid facing the fact that we are all solitary and
perishable.

Jacques Cousteau

JACQUES COUSTEAU MET Simone Melchior at her family's Paris
apartment in the summer of 1936. She was seventeen, the daughter
and granddaughter of French admirals, born on January 19, 1919, in
Oran, Algeria. Until she was five years old, her family lived in
Toulon, the home port of the Mediterranean fleet, then moved to
Japan, where her father was a diplomatic attaché. By the time the
Melchiors returned to France, Simone was a teenager fluent in Japanese and infected with wanderlust. Though her horizons were limited by her gender, she was a lycée student whose interests went far
beyond keeping a house. She would later say that her dreams always
included the sea.

The soirée at which Cousteau met Simone was one of many
hosted by Marguerite Melchior to introduce her daughter to eligible
navy men. Cousteau was fixed on Simone from the moment he
walked in the door. Feigning indifference, he registered a stunning,
compact young woman with flaxen hair, high cheekbones, and a
seductive mélange of amusement and confidence in her alert green
eyes. For her part, Simone quickly picked out the animated, wiry
young man holding a movie camera of all things, not something she
expected at one of her mother's parties. He was obviously navy, judg-
ing by his posture, but dressed in civilian clothes. She was instantly
enchanted by his presence, a blend of curiosity, delight, and confidence, and she liked the way he brashly panned his camera around the
roomful of people holding cocktail glasses and chattering. He had a
long, rural face, deeply etched by lines in his forehead pointing down

to a long, sharp nose. Another set of concentric arcs wrinkled each side of his mouth, emphasizing a dazzling smile that he seemed to measure out for effect. As he wound the camera with its little silver handle, he moved his arms as though they hurt him. Simone's father easily noticed his daughter's interest in the officer with the camera, led him across the room to her, and introduced Ensign Jacques-Yves Cousteau. Admiral Melchior added that Cousteau was an aviation cadet living in Paris with his family while recuperating from injuries suffered in an automobile accident.

Cousteau took over the conversation, explaining the accident to Simone as though telling an adventure story in which he was simply a character swept along by events beyond his control. Five months earlier, he was just about to graduate from flight school when he borrowed his father's Salmson sports car to go to a friend's wedding in the hills west of Bordeaux, where he was based. The Salmson was fast and nimble, but when the headlights suddenly dimmed on a hairpin curve, he careered off the road into the darkness and blacked out. Only the luck of a passing farm truck on the otherwise deserted mountain road saved his life. Cousteau regained consciousness in the

Simone Melchior Cousteau
(PRIVATE COLLECTION)

hospital with twelve broken bones and a paralyzed right arm. His doctor told him that his arm was numb because it was infected and probably should be amputated. Cousteau refused amputation. He would rather die than live as a one-armed man. A few days later, the infection resolved itself and his bones began to heal. After months of physical therapy, Cousteau was out of plaster and walking around, but he was in constant pain. He was finished as an aviator, and though he could not know it at the time he was again very lucky. Every man in his flying class would be killed during the first few weeks of the war with Germany, which was three years away.

Cousteau, at twenty-six, was nine years older than Simone, and he had already sailed around the world—with his camera—on the training ship *Jeanne d'Arc*. Flying airplanes had been one of three possible life courses Cousteau had ceremoniously announced to his family as a boy. His other two choices were being a film director or becoming a medical doctor. Now, he was not at all sure of his future. Cousteau had been a curious but anxious boy who had become a man in love with impulse and driving fast cars on mountain roads. The rural face Simone had noticed right away was a true compass to his beginnings. He was born in St.-André-de-Cubzac, an ancient village at the confluence of the Dordogne and Garonne rivers, which form the Gironde Estuary, which flows into the Atlantic Ocean seventy miles to the northwest. The first traces of habitation of the fertile, temperate hillsides are shards and bone fragments more than ten thousand years old, evidence of seasonal encampments of nomadic bands foraging in the north after the last ice age. The Pax Romana had reached the Gironde two thousand years ago, creating the city now called Bordeaux with the profits from sugar, tobacco, indigo, cotton, ebony, cocoa, coffee, slaves, and oysters exported to the rest of the Roman Empire. When the wild oysters grew scarce, the Romans figured out how to raise them in sea farms from spat on discarded shells, a method still unchanged by time. Soon after they arrived, the Romans planted grapevines and made wine that would forever after define the place. The English took over Bordeaux in the twelfth century when Louis VII's divorced wife, Eleanor of Aquitaine, married the Plantagenet king Henry II. Their kin ruled the region for three hundred years, as the wines of the magnificent, fertile valleys—called Bordeaux by the French and claret by the English—flowed by the shipload from convenient docks along the Gironde to London and the rest of the British Empire.

The vineyards nearest St.-André-de-Cubzac are known for wines made from common grapes that are perfectly drinkable and preferred by locals because they aren't as expensive as the vintages of nearby St.-Emilion and Poulliac and the others that find their way into the cellars of oenophiles. Cousteau's mother, Elizabeth Duranthon, was one of five daughters in St.-André-de-Cubzac's wealthiest family, descended from generations of landholders and wine merchants who had tended those vineyards as though bound to them by inherited vows. His father, Daniel, was a lawyer, one of five sons of the village *notaire* who was responsible for witnessing deeds, marriage contracts, and other consequential transactions. Daniel was the brightest of the Cousteau sons, so the family had marshaled its merchant-class resources to send him through law school in Paris. He returned to his home village of 3,800 as a fully fledged *notaire* himself, but remained for only three years, executing documents in his father's office and, not incidentally, courting Elizabeth. She agreed to marry him in the winter of 1905. The following spring, thirty-year-old Daniel and eighteen-year-old Elizabeth Cousteau boarded a train for the long day's journey to Paris; thereafter they rarely went back to St.-André-de-Cubzac.

In Paris, Daniel Cousteau had only one client, an American expatriate his own age named James Hazen Hyde who had inherited a fortune from his father, the founder of the Equitable Life Assurance Society. Hyde had fled the United States under a cloud of controversy and allegations that he had cooked the books of the insurance company after he took control. He sold out, moved to France, and lived royally, which included hiring Daniel as a legal adviser, traveling companion, private secretary, and tennis partner. Daniel was essentially a servant, though he was part of Hyde's inner social circle, and loved the life of leisure and pleasure. Daniel Cousteau and his wife were gilded vagabonds with no real home, but reckoned their life was a far better adventure than notarizing deeds in a quiet village or taking a minor position in her family's wine business. He was a man of only average intelligence who possessed finely tuned primitive instincts for simply getting things done. He had realized early in life that his wits and careful decisions were his only sources of real wealth, and he took advantage of opportunity as naturally as a fox snatches a chicken. Daniel was well aware that the only estate he would likely leave to whatever children he and Elizabeth might have was his example as a skillful liver of life.

Daniel and Elizabeth Cousteau's first son, Pierre-Antoine, was born in Paris in December 1906. A week later, Daniel headed south for the winter aboard James Hyde's yacht. Elizabeth returned to St.-André-de-Cubzac to show off the baby to her family, after which she joined her husband and traveled with Hyde's entourage for four more years. Though the high life continued to appeal to the Cousteaus, the constant traveling and frequent separations took their toll on Elizabeth and Pierre. They had their own apartment in Paris but were rarely there. They returned home to the Gironde only for funerals and weddings. In the spring of 1910, Elizabeth was pregnant with her second child and the Cousteaus decided that she should return to St.-André to have the baby with the help of her family. On June 10, 1910, the boy they named Jacques-Yves arrived in an upper bedroom of the Duranthon house across a courtyard from the village church. Though the Cousteaus were only nominally Catholic, Daniel returned to St.-André for the baptism of their second son.

For three more years, the indulgences and adventures of James Hyde continued to determine the patterns of the Cousteau family. "My first conscious memory was of swaying in a hammock in a railroad coach as the train steamed through the night," Cousteau later said. "It was all okay until the war."

In September 1914, when Jacques-Yves was four years old, the German army reached the Marne on the outskirts of Paris. From there, the bloody western front extended along the river from Verdun to the English Channel. The French government retreated to Bordeaux, the lush life in Paris shriveled to one of subsistence, and James Hyde left Europe to lie low in America until things quieted down. He reluctantly let Daniel go in the winter of 1914, forcing the Cousteaus to survive for the next four years on Elizabeth's family money. Jacques-Yves developed chronic enteritis and a cascade of stomach inflammations, which added to the family's wartime woes and marked him as a sickly child. The plump, smiling infant in the earliest family photographs became a gaunt, earnest little boy with the guarded expression of a child who didn't feel very well and wondered what was going to happen next.

The Cousteaus stayed in Paris until they had trouble finding enough to eat, and then went back to St.-André, where at least the family gardens could feed them. By the time Jacques was seven, he had become a quiet little boy with no interest in games or sports, uncomfortable with children his own age. He wasn't defiant, but seemed indifferent to most of what happened around him. His parents were encouraged only by his apparent attraction to mechanical things. He could spend hours using tree branches to move roots and rocks around in the garden.

When the war ended in 1918, a new generation of rich and restless Americans migrated to Paris, where they could live well on inflated dollars against nearly worthless postwar French francs. Among them was a bon vivant named Eugene Higgins, the son of a carpet manufacturer who had inherited $50 million when his father died in 1890. Middle-aged in 1919, Higgins had never married but was seen with the most glittering women in New York and Europe. He spent his time and money perfecting his hunting, horseback riding, fishing, golfing, and boating, and throwing extravagant parties. At one of his galas in Paris, Higgins was introduced to Daniel Cousteau, who, thanks to his years with James Hyde, still circulated among well-off expatriates. After a cordial first conversation and a week of checking references, Higgins offered Daniel a job as his factotum. There was only one catch. Higgins owned the house in Paris, a brownstone on Fifth Avenue in New York, and an estate in the rolling hills of New Jersey across the Hudson. Daniel and his family would have to move to New York for part of each year when Higgins joined the social season in Manhattan.

In early 1920, the Cousteaus sailed for the United States aboard a French Line ship. Jacques had never been happier in his life than he was during the eight-day voyage. He explored every passageway and compartment aboard the liner and became a pet to stewards, engineers, and deck officers who were happy to show him around. Daniel and Elizabeth were stunned by how gregarious their timid son was on the ship, finally showing an interest in something other than tinkering quietly by himself. Previously, they had been genuinely worried that he was doomed to be a frail, unhealthy loner.

In New York, the Cousteaus settled into an apartment on West Ninety-fifth Street that would be their home for two years. Daniel traveled with Higgins, but Elizabeth stayed put, insisting that their sons needed something resembling a normal childhood to have any

chance for success in life. Fourteen-year-old Pierre-Antoine, by then
known as PAC, went to DeWitt High School. Ten-year-old Jacques,
who was calling himself Jack because it sounded more American and,
not incidentally, rhymed with PAC, went to Holy Name School.
Both boys struggled at first with conversational English but were
finally comfortable enough in the new language to appreciate that
American schools were much more relaxed than the rigid academies
they were used to in France. Jack continued to bond to his older
brother as a safe haven in a sea of American boys who never completely accepted him because of his accent and his social reticence.
PAC was Jack's single friend, creating the enduring sense that only a
member of the family could ever be fully trusted.

Eugene Higgins demanded that Daniel Cousteau compete well
with him at tennis and golf, and shadow him among his residences, a
yacht, and the resorts he favored as the seasons changed in Saratoga
Springs, Newport, and Deauville. Those absences created a pattern of
serial abandonment that forced Daniel's sons to adjust to the strict but
feminine nature of their mother. They lived with the unspoken fear
that displeasing her might result in the unthinkable: abandonment by
both parents. For Jack especially, the presence of a woman became as
essential to survival as oxygen. Without her, he and PAC would have
been alone in a foreign country. The thought terrified him.

Despite Daniel's comings and goings, the Cousteau brothers and
their mother found a comfortable routine in New York. PAC, who
was stocky and muscular, thrived on playground games with the other
boys, and his life began to ripen away from the apartment on Ninety-
fifth Street. Jack was more gregarious than he had been in France, but
he still conveyed the image of a frail child, and being around other
boys was hard for him. He wasn't good at hitting and throwing balls.
He didn't like games of chase that left him winded and always the first
one tagged out. He hated boxing, which was one of the highlights of
the gym class at school. Part of it was his still-fragile constitution. Part
was that he just didn't want to spend time on what did not interest
him. As a student, he barely kept up. What Jack did like was building
things. Stacks of blocks had led to wooden levers in the garden, which
became wood-and-glue models of boats, planes, and machines. Most
of the time, after a ceremonial presentation to PAC and his mother,
Jack's creations languished in his bedroom, neatly arranged on shelves

he had built himself. During his second winter in New York, when he was eleven, he built a working scale model of a dock crane from a set of plans published in *Popular Mechanics* magazine. The crane was as tall as he was, and could be controlled by cranks and pulleys to lift, swivel, and move forward and backward on rolling cams he added to the design. The next time his father was in New York, Jack demonstrated the crane for him in their apartment living room. Later, Daniel described the crane to an engineer working on one of Higgins's yachts, who told him that moving the crane with the rolling cams would be a genuine improvement in the real crane. If it wasn't on the blueprints, Jack could probably apply for a patent. Daniel and Elizabeth didn't push Jack to file a patent for the crane, but they were very relieved that he was finding a passion in building models that might translate into a productive career as an adult.

Jack still seemed shy, reluctant to assert himself with other children, and fearful of rejection, so his parents sent him to summer camp on a Vermont lake in hopes of improving his social life. It was a two-week endurance test of tightly scheduled days filled with crafts, forest lore, swimming, canoeing, and horseback riding. Jack handled most of it, but he hated the horses. On the first day, Jack's horse threw him; rather than sympathizing with the frightened little boy, the riding instructor, a German émigré named Boetz, scolded him and threw him back in the saddle. With Boetz and the other boys glaring at him as though he was a pathetic failure at an essential manly duty, Jack glared back, got off the horse, and refused to get back on. As punishment, Boetz ordered Jack out of the class and sent him to clear dead tree branches from the camp's swimming pond, a chore no one else wanted to do because groping around underwater was as frightening as entering a haunted house. For Jack, submerging into the brown, silty lake was bliss. He opened his eyes underwater for the first time in his life, and even though he could see only a few inches in front of his face, he was relieved to be insulated from the world above with its horses, Boetz, and a pack of jeering boys. For the rest of his life, Cousteau would tell the story. He was not at all frightened as he methodically dove to grab the waterlogged limbs, surfaced, and swam them across the shallows to shore. The water soothed him and banished all fear. "I loved touching water," Cousteau said. "Physically. Sensually. Water fascinated me."

By 1923, when Cousteau was twelve years old, Paris had regained some of its elegance and vitality, at least for the upper class. Eugene Higgins packed up his mistress, the Cousteau family, and three other servants and moved back to France. Elizabeth dug in again and refused to travel with the entourage, preferring the placid routines of guiding her sons through their lycée years from their Paris apartment. Soon, however, PAC talked his father into allowing him to quit high school a year early, skip university, get his military service out of the way during peacetime, and go to work making money. Because the Cousteaus lived in the rarefied atmosphere of wealthy expatriates, politics were of little interest to them, even when billions of francs of war debt collapsed the French economy. Riotous collisions between the left and right dominated newspapers, with each side promoting its own vision of a hopeful future for the republic. Most people searched for solutions to the chaos of France by focusing on Stalin and his Communist experiment in the Soviet Union or on Mussolini and his fascists in Italy. Daniel, Elizabeth, and Jacques barely noticed any of it. PAC alone dove into the political fray like a man hungry for a fight, becoming an opinionated leftist while he performed his military service for eighteen months. Initially, he argued for reforms to benefit the working class, quick to point out that a leftist or a worker had no business in any army that might demand his life in defense of what, to him, was a corrupt, inhumane society.

When his brother left home, Jacques missed him terribly. He told everyone that he would now like to be known as JYC, pronounced *Jeek,* as Pierre-Antoine had become PAC. Without his brother, it became even harder for JYC to connect with the outside world. He had crossed the Atlantic Ocean twice, lived in New York City, and traveled Europe with his family, but venturing into the unknown world of other kids who had never been out of Paris was frightening to him. He was desperate to be liked, but had none of the tools for becoming popular. He still hated sports. He was a poor student. He had not deciphered the codes of small talk. Most of all, he had become a gangly boy with a large nose and a long face who knew that by no stretch of the imagination could he be considered attractive to girls. JYC sleepwalked through his days at school, kept his mother company when his father was away, and spent as much time as he could in the dependable haven of his bedroom workshop. Alone, he tinkered with his models and read about the flood of interesting machines, amusements, and technology that had inundated newspapers and magazines as the world exploded with new inventions during the Roaring Twenties.

In the summer of his fourteenth year, JYC came across an article about the Pathé brothers, who owned the patents for the original movie cameras that had been invented by Auguste and Louis Lumière. It was impossible for anyone living in Paris, even a reclusive teenager, to be unaware of the French dominance of the sensational new art of the cinema. The Lumière brothers' film *L'arrivée d'un train en gare de La Ciotat* (*A Train Arrives at the Station*) had launched the revolution in 1895, and in less than a decade the technology of moving pictures had spread around the world. JYC knew all about the Lumières, and also idolized Georges Méliès, who made the first science fiction film, *Le voyage dans la lune* (*A Trip to the Moon*), in 1902. When he saw it twenty years later, the silent classic that ended with the astounding image of a rocket ship stuck in the eye of the man in the moon enchanted JYC. He learned that Méliès was a stage magician before he was a filmmaker, and had invented cinema's first special effect by accident. While he was shooting a man walking, his camera skipped a few frames. When he looked at the developed film, the man disappeared, then reappeared. The next day, Méliès stopped his camera on purpose as he shot a scene of a magician and a woman assistant. The magician waved his wand; the assistant disappeared. Méliès called his effect the stop trick and used it in his 1907 adaptation of Jules Verne's *Vingt mille lieues sous les mers* (*20,000 Leagues Under the Sea*), another film that spoke directly to JYC in his adolescent seclusion.

By 1924, the Pathé brothers had transformed their gramophone shop in Paris into a motion picture supply house cluttered with used equipment and workbenches from which innovation flowed constantly. After one visit to the shop and three months of saving part of the allowance his mother gave him each week, JYC bought a used Pathé Baby, a hand-cranked camera that exposed sixteen frames of 9 mm film through a 20 mm lens. Two turns of the handle produced one second of running time. JYC bought the camera mostly because he wanted to take it apart to see how it worked, and he spent hours dismantling and assembling it before shooting his first roll of film. The chemicals and techniques of film processing also fascinated him. With a metal tank and jars of developer, stop bath, and fixer he bought at Pathé Brothers, he turned a minute of panning around the street outside his house into a sequence on a strip of cellulose that, incredibly, reproduced motion. A month after buying the Pathé Baby, JYC shot his first full movie of a cousin's wedding. He was nervous when he developed the film because all those people at the wedding had seen him with his camera. If he failed, they would all know it. When he uncoiled the last of the rolls from the developing tank and saw that it was as perfect as the other two, it was the most satisfying moment of his life to that point. With instructions from one of his cinema magazines, JYC cut and spliced his movie with special tape he brought at Pathé, then rolled it in a continuous strip onto a large reel. When the bride and groom returned from their honeymoon, he exhibited his incredible record of their wedding day with a hand-cranked projector in a room full of relatives, including his mother. It was three minutes of grainy images shot in a church and at the party following the ceremony. People waved, smiled, and did little dances, obviously delighted and impressed by the fourteen-year-old boy dipping and turning to record that moment in their lives. After the last images flickered through the projector's gate to be replaced by the glare of the unshielded lightbulb, the little crowd burst into applause. JYC was overwhelmed by the reception his first movie received but more so by the astonishing privilege he had enjoyed at the wedding itself. He could go anywhere he wanted to go. Everyone wanted their pictures taken. With the camera, he was popular.

JYC's films soon had plots, villains, and dramatic effects borrowed from the cinema masters of the times. He mounted his camera on a moving car to give the point of view of a passenger, as E. A. Dupont had done in his film of a trapeze artist in *Varieties*. Cousteau featured himself in front of the camera in costume and false mustache as a hero or villain, developing an ease with himself as the center of attention. In one of them, the camera tracks an open roadster coming to a stop. The young man driving looks exasperated and gets out to see what's wrong, leaving a girl who was sitting beside him in the car. JYC enters from out of the frame, mugs at the camera, which is now in someone else's hands, and jumps over the door into the car. He grins deviously in a close-up before a hand reaches into the car and, using Georges Méliès's stop trick, yanks JYC back over the door. Another of Cousteau's earliest films begins with the camera in the hands of someone other than him, a scene in which JYC and a pretty girl are arguing with exaggerated gestures in a rowboat on a lake. They stand up, she pushes him, and he goes over the side into the water. After a sidelong glance at the camera, JYC sinks beneath the surface and the words "The End" appear. JYC taught himself how to shoot titles and credits on signboards listing J. Cousteau as producer, director, cameraman, and actor. He named his production company Société Zix.

Even with his camera and a troupe of players waiting for him when he finished school every day, JYC had no patience for being pent up in a classroom and very little interest in what his teachers were telling him about mathematics, literature, art, and culture. In the spring of 1928, his grades were so bad that Elizabeth took away his movie camera to punish him. A week later, JYC broke some windows in one of the stairwells at school. He was caught in the act but insisted that he was conducting a scientific experiment to determine if a weakly thrown stone makes a bigger hole than a strongly thrown stone. The headmaster suspended him for a week, and JYC's parents had had enough of their rebellious, self-contained son.

After consulting with Daniel during one of her husband's brief interludes at home, Elizabeth dispatched JYC to a boarding school in Alsace-Lorraine, 250 miles from Paris. Without his camera, and with none of the new friends he had made as a filmmaker, he somehow thrived. The school was full of unruly boys sent there for discipline, and the harsh treatment of their teachers and overseers created a camaraderie among them that inspired JYC. With the assumption that he was going to be criticized and punished no matter what he did, he was free to excel on his own terms.

After graduating at the top of his class in the spring of 1929, he returned to Paris for the summer but knew he couldn't live at home for long. His prospects for making a living as a cinema director were dim, so he began thinking of making movies as a hobby. Instead, he chose the second of his childhood ambitions, to fly airplanes in the navy, and took the exams for admission to the French naval academy in Brittany. No one was more surprised than he was when he passed. Life at sea itself did not appeal to him, but he had vivid memories of his voyages to and from America and knew ships could connect him to the wonders of the world outside France. Four years later, after a globe-circling cruise aboard a training ship, he received his commission in the French navy. In January 1936, just before receiving his wings as a pilot, he drove his father's sports car off a mountain road. Within hours of regaining consciousness in a hospital, Cousteau knew his career as an aviator was over.

While he recuperated, uncertain of his future in the navy, Cousteau returned to his vision of himself as a filmmaker. He still liked the way his camera set him apart from the rest of the world and gave him permission to look right at people and things around him. It was a novelty. It was almost magic. He was in control. He had the navy as a career but could not bring himself to set aside storytelling with his camera.

In his film of the soirée where he met Simone Melchior in Paris in 1936, she looks directly at the lens and clearly mouths her name, "Simone." She exaggerates a coy smile with downcast eyes, the gesture of an actress admitting that she is acting in Cousteau's movie. The camera pans away from her, then returns to find her beaming as her natural self. Cousteau and Simone became a bonded pair at that moment. The navy and the sea were common to them, and they recognized something in each other that made them feel safe. Both had been children of absent fathers, both had traveled while most people of their generation were living sedentary, predictable lives. They saw in each other a certain delight for their surroundings, as though their presence itself transformed an ordinary parlor into a movie set. Their first conversation, after Cousteau dropped his camera to his side, was about boats. For a year after they met, they were as inseparable as a man and a woman living 500 miles apart could be. JYC shuttled between the navy base in Toulon and Paris while remaining on light duty. Simone finished high school. A month after her graduation, on July 12, 1937, they were married at St.-Louis-des-Invalides, honeymooned in the Alps, and returned to the fleet in Toulon.

2

LES MOUSQUEMERS

IN TOULON, Cousteau spent some days training with other sailors and some days at the hospital enduring painful therapy for the wounds he suffered in the car wreck. The navy had decided to make him a gunnery officer who would eventually be an expert on munitions in charge of a gun crew. He tired easily, rejected painkillers except when he needed them to sleep, and had trouble keeping up during the most ordinary of days. In the fall of 1936, while he was aboard the cruiser *Condorcet,* one of the other officers drew him aside and gently suggested that there might be a better way to regain the strength in his arms than what the doctors had prescribed. Philippe Tailliez, who was slightly older than Cousteau, seemed shy and spoke with a stammer, but he was obviously a quietly confident officer, respected by his men. Swimming in the sea, Tailliez proposed, would help him build his strength. Cousteau told Tailliez that he knew how to swim, loved the water, and was willing to try anything that could relieve the pain and get him back to full duty. The following day, Tailliez took Cousteau to a rocky beach below the bluffs of Sanary-sur-Mer just west of Toulon.

Tailliez was anything but shy in the still-warm water of the Mediterranean in autumn. At thirty-one, he was a raw-boned, agile man whose placid exterior concealed boundless energy and a passion for the sea that had been kindled during his childhood on Dunkerque beach in northern France, where he was born in 1905. Philippe's father was a sailor who returned home from his voyages with tales of the South Pacific, pearl divers, and underwater fishermen who hunted with spears. Philippe's father also introduced his son to the mortal danger of the sea with a story about a child diving for coins near a warship anchored in the Suez Canal. The little boy, no older than Philippe's own seven years, waved to the French sailors, disappeared beneath the surface after the coins, and bobbed up dead a few minutes later.

Les Mousquemers, *1948*. (Left to right) *Jacques Cousteau, Philippe Tailliez,
and Frédéric Dumas* (COURTESY OF WWW.PHILIPPE.TAILLIEZ.NET)

When the Tailliez family was living at the navy base in Brest,
Philippe had won a goldfish in a glass bowl by spinning a prize wheel
at a charity bazaar. On the way home, the rocking of the car sloshed
the water around in its jar and Philippe watched the fish stabilize itself
against the motion with subtle, perfect adjustments of its tail. The
next time he went swimming, Philippe added a fishlike kick to his
breaststroke and found that it increased his speed. He began to claim
the ocean as part of his natural habitat. The cold, cloudy waters of the
English Channel and the Atlantic were hospitable only to tentative
underwater excursions, but when Philippe was posted to Toulon after
completing his officer training at Brest, the clear, warm Mediterranean turned out to be paradise.

With instructions from the commander of *Condorcet,* a skin-diving
enthusiast named Captaine Louis de Corlieu, Tailliez made himself a
pair of flippers by sandwiching pieces of metal saw blades between
two slabs of rubber and strapping them to his feet with twine. The
fins, about a foot long, more than doubled the power of his kicks. Tailliez also fashioned a mask from aviator glasses. Like eyes, which have
evolved independently in many different kinds of animals, goggles for
seeing underwater had appeared many times on many seashores all
over the world. The earliest versions were double rubber or fabric cups

with small glass panes, held in place by a head band. Like every skin diver, he continued to experiment.

Tailliez figured out how to strap a few pounds of lead weights to his waist so he didn't have to struggle against his own buoyancy. He also cut himself a two-foot length of garden hose tied with twine into the shape of a J, through which he could breathe on the surface while scanning the water below him for prey. To dive, he held his breath. When he surfaced, he blew the water out of the snorkel tube. His ultimate piece of equipment was four feet of solid brass curtain rod, one end sharpened into a point, the other taped to a loop of automobile inner tube. The diver could draw his spear back against the elasticity of the rubber band on his wrist, and fire it with enough force to pierce a fish or lobster a few feet away when he released the tension. Tailliez was eating well, but sometimes left his spear on the beach and just cruised the surface, marveling at the world beneath the sea, which seemed utterly private and his alone.

In the autumn of 1936, Cousteau and Tailliez ended most days in the water at Le Mourillon Bay near Sanary-sur-Mer, Tailliez hunting while Cousteau exercised his sore arms with a gentle crawl stroke on the surface. When Tailliez was successful, which was more often than not, they would build a fire on the beach, roast one of the fish, and eat it while shivering over the embers as dusk fell. Usually, Tailliez killed enough fish to pass around in Sanary, where he was a bit of a hero because of his generosity. Soon, the two men had established the routine from which they strayed only when they were on duty or when the water was simply too cold.

One afternoon, Tailliez was preparing for his hunt when he casually offered his mask, fins, and snorkel to Cousteau, who protested that he was a navy gunner interested only in perfecting his crawl style in case he had to abandon ship someday. The sea, he insisted, was merely a salty obstacle that burned his eyes or a way to travel from port to port. Tailliez coaxed. Cousteau gave in. The older man helped his friend into the gear and swam with him as they eased into the Mediterranean. Cousteau listened to the rasp of his own breath through the hose, paddled over the seascape in the brightly illuminated shallows, and saw it all clearly through his goggles. Minutes after entering the water with Tailliez's mask, fins, and snorkel, Cousteau's life changed forever. Below him through the crystalline water he saw rocks covered with green, brown, and silver forests of algae and brightly colored little fish he never knew existed. He stood up to breathe and glanced on shore, where he saw a trolley car, people, and electric-light poles. Another world entirely. He put his eyes under again and civilization vanished. Cousteau was in a jungle never seen by those who floated on the opaque roof of the sea. He remembered Mr. Boetz, horses, and the feeling of bliss that he had discovered in the silty brown water of a Vermont lake. Now, instead of groping blindly for branches in the murk, he could see clearly.

Shepherded by Tailliez, Cousteau ventured into deeper water. He was startled to see that he could pick out details on the bottom more than 20 feet below. The fractured talus of the beach slid into underwater shoals of rock and sand alive with ochre fronds of kelp and patches of bright green sea grass. Bright constellations of red, orange, and lavender starfish decorated every surface to which they could cling against the surge of the sea. In the crevices of rock piles and eroded boulders, he picked out the dark purple spines of urchins. Swarms of fish were weightless and agile, with repertoires of acrobatic stunts and bursts of speed Cousteau had never seen on land. Lumbering groupers hovered like dirigibles near the larger rocks, disappearing in a single heartbeat when he startled them. Schools of synchronized mackerel flashed in the sunlight as though commanded by an invisible choreographer. A codlike merou poked at the bottom, and a solitary palomata that looked as if it weighed 200 pounds lurked in the distance. Cousteau wondered about the size of the palomata and held his hand in front of his mask to determine the scale. He realized for the first time that the entire sea through glass was a magnifying lens. Cousteau held his breath, dove down 15 feet to a boulder, held on to it, and spotted Tailliez splashing above.

"The reason I love the sea I cannot explain," Cousteau said. "I only know that sometimes we are lucky enough to know that our lives have been changed, to discard the old, embrace the new, and run headlong down an immutable course. It happened to me on that summer's day, when my eyes were opened on the sea."

Cousteau began spending hours in the library at the navy base reading about the sea and swimming underwater. "Water is H_2O, hydrogen two parts, oxygen one part," D. H. Lawrence had just written, "but there is a third thing that makes it water and nobody knows what that is." Water covers 71 percent of the earth's surface with 328 million cubic miles or 361,200,000,000,000,000,000 gallons. Water is essentially colorless, but its density and composition scatter the electromagnetic wave lengths of red, orange, yellow, and green light. It absorbs every color but blue, creating the dominant hue seen from a beach or from the moon. The phenomenon can be observed by a deep diver descending into the abyss as the light deteriorates in the order of the spectrum beginning with red until, at 600 feet, only the blue remains. At about 1,900 feet, the radiation emitted by the sun has been slowed enough by its passage through water that it has lost its power to penetrate the human retina. Beyond, the ocean is black. The average depth of the ocean is 2.5 miles, with abyssal chasms reaching 7 miles.

Seawater is never just water. Oxygen and hydrogen combined into water molecules make up 96.5 percent of the ocean, but the other 3.5 percent consists of dissolved elements. Every cubic mile of the ocean contains 90 million tons of chlorine, 6 million tons of magnesium, 4 million tons of sulfur, almost 2 million tons each of calcium and potassium, 132,000 tons of carbon, 2,350 tons of nitrogen, 50 million tons of salt, and 38 pounds of gold. The weight of seawater varies depending upon how much salt and other elements it carries, but it averages about 8.35 pounds per gallon. The salt content of seawater is about 34.7 parts per thousand, roughly the same proportion as that found in human blood and amniotic fluid. In fact, the embryos of mammals, including human beings, pass through a developmental stage during which they have vestigial gills.

The molecular structure of water is incredibly stable. Once two hydrogen atoms fuse with one oxygen atom, the bond is almost indestructible, whether the water is a liquid, a gas, or a solid. And unlike other molecules, which require applications of enormous amounts of energy to change their states, water demonstrates its versatility under ordinary temperatures and pressures. Water can exist in any of its three states under conditions that tolerate and even encourage plant and animal life.

Though no one has come up with a definite explanation for the origin of the oceans, it is certain that life began in the saline soup of the sea. The first traces of life on the 4.5-billion-year-old earth are the fossils of cyanobacteria, which appeared about 3.5 billion years ago. Single-cell organisms called eukaryotes appeared 1.5 billion years ago, taking another step in the direction of complexity, with nuclei able to pass on genetic information to succeeding generations. About a billion years ago, multicellular animals and plants appear, but life did not leave the sea until just 400 million years ago. The process of creating life in the sea depends upon drawing energy from the sun to build carbon compounds from carbon dioxide and water in a process called photosynthesis. Life is also sustained in the darkness of the abyss far from sunlight by the production of organic molecules in a process called chemosynthesis, which depends upon the interaction of bacteria with sulfur, methane, and other chemicals. A by-product of both types of syntheses is oxygen.

There are more detailed maps of the surfaces of the moon, Venus, and Mars than of the ocean floor, less than 5 percent of which has been glimpsed by human eyes. People have been gazing with awe and wonder at the ocean since consciousness evolved to allow such sensations, and at some point, a brave or foolish hominid stuck his head beneath the surface to see what was there. The obvious conclusion of that first excursion was that humans cannot breathe underwater, and that we can stay under only for as long as we can hold a breath of air in our lungs, about two minutes for an average person. For most of human history, the sea beneath the surface was a mysterious realm inhabited by creatures with frightening teeth, beaks, and tentacles. It was a tomb for unlucky mariners, a place hidden from us because we breathe air. Obviously, diving in the ocean to reach food or to work beyond the shallows meant holding one's breath or finding a way to bring air from the surface.

In one of the books he read, Cousteau came across a two-thousand-year-old sketch of a naval battle between the fleets of Greece and Syracuse that depicted saboteurs swimming underwater, breathing through reeds to drill holes in enemy ships. Another ancient image showed Alexander the Great on the seafloor in an overturned barrel, breathing the air trapped inside. At the end of the seventeenth century, Edmond Halley, better known for his discovery of the comet that bears his name, invented a weighted wooden box with a glass top in which he could descend for a few minutes to about sixty feet. He breathed air from skin bladders lowered to him on a rope. Jules Verne borrowed Halley's air bag method in *20,000 Leagues Under the Sea*. The fictional diver from the submarine *Nautilus* walks around on the bottom of the ocean towing a balloon of air behind him. A hundred and fifty years after Halley, divers really were walking on the floor of the ocean, sustained by air fed through hoses from hand or powered pumps on the surface. Their depth was limited to 60 feet or so, and though they were able to work underwater as salvors and mechanics, their weighted boots, heavy copper helmets, and bulky suits made hunting difficult except for harvesting abalone, sponges, or other sessile prey.

After that first day wearing Tailliez's goggles, only his time with Simone was more important to Cousteau than skin diving. The next time he went to Le Mourillon, he brought his own mask, fins, weights, snorkel, and sling spear to hunt in the jungle below. For a man whose primitive instincts were paramount in his relationship with the world around him, he was entering paradise. He quickly learned to adjust his aim to compensate for the distortion of the water. His ability to react to movement in the periphery of his vision would reward him with a kill. He was acutely aware of his sense of direction and balance in the weightless world below the surface, and easily imagined himself to be a fish stalking prey.

"Soon, I listened to gossip about heroes of the Mediterranean, with their Fernez goggles, Le Corlieu foot fins, and barbarous weapons to slay fish beneath the waves," he wrote. "I was obsessed because hunting underwater suited me so well." Cousteau quickly intuited that free diving to hunt was only the beginning of his life underwater.

By the following summer, only the slightest twinges of pain remained in his arms. He and Simone had settled in Sanary-sur-Mer, six miles from the base at Toulon and three miles from the neighboring village of Bandol. Sanary was an officers' enclave and, coincidentally, home to a community of expatriate German intellectuals that was growing steadily as Hitler imposed restrictions on academic freedom and expelled Jews from universities. To the west, in Spain, civil war ruptured any hope for continued peace there. To the southeast, Mussolini was barking about Albania, which he said rightfully belonged to Italy.

In the autumn of 1937, JYC and Simone opened the doors of their cottage to Tailliez and other officers, enjoying the steady flow of people through their seaside home. PAC; his wife, Fernande; and their infant daughter, Françoise, visited from Paris. After PAC's service in the demoralized, chaotic French army between the wars, he had craved order so much that he abandoned his leftist politics and embarked on a career as a fascist, anti-Semitic journalist. He was certain that only a rigidly controlled population could ever resurrect Europe from the dredges of constant warfare and economic domination by Jews. PAC risked alienation from his family by calling his father's decision to let him quit high school to join the army an example of "deplorable liberalism." PAC, however, didn't make the mistake of insisting that his brother share his politics. JYC and Simone agreed with some of what PAC brought to their dinner table in Sanary but for the most part tried to steer the conversations to less flammable topics.

Skin diving was at the top of the list. Most of all, the Cousteaus and their guests—whether the radicalized PAC, German expats, or other navy men—enjoyed the thrill of swimming and diving together in the sea. Tailliez remained a passionate hunter, but JYC turned most of his attention to designing and building a housing for an underwater camera. Through the winter, most of the Cousteau family circle treated JYC's obsession with making moving pictures underwater as a fantasy. By the time the sea warmed up enough for diving the following spring, he was ready to test his first waterproof camera. At the shallow depths of breath-hold diving, his main problem was simply keeping the camera dry. He knew that if he went too deep, water pressure would become an issue, but for his first camera he ignored it. He bought a used 8 mm Beaulieu, braced it on a bracket he built inside a gallon fruit jar, and set its timer to trigger automatically to record thirty seconds of action. It was absurdly simple. The glass of the fruit jar was nowhere near lens quality, but the contraption worked. After his first dive with it to a depth of 20 feet, he huddled in his darkened bathroom, carefully wound the film onto the reel of a developing tank, and added the usual succession of chemicals. After what seemed like an eternal half hour of pouring, agitating, and draining, he held the dripping-wet film in front of a lightbulb. There, in a sequence of blurry images depicting motion, was Simone splashing on the surface against the glare of a sunny sky. He wasn't sure

whether anyone else on earth had ever shot a motion picture underwater, but he had proved to himself that it could be done.

Cousteau's second roll of underwater film produced a few jerky
frames of a skin diver swimming directly at the camera, grinning like
the happiest man alive. The diver was Frédéric Dumas, a twenty-five-
year-old civilian, the son of a physicist, a champion swimmer, free
diver, and spearfisher. Everyone on the Riviera knew him as Didi. He
seemed to exist with no visible means of support, spending part of
each year simply living on the beaches of Le Mourillon Bay. He
insinuated himself into the Cousteau household in the early spring of
1938 after watching Tailliez hunting offshore.

"One day, I am out on the rocks," Dumas told the Cousteaus on
the first day they invited him to lunch, "and I see a man much further
on in evolution than me. He never lifts his head to breathe, and after a
surface dive water spouts out of a tube he has in his mouth. I am
amazed to see rubber fins on his feet. I sit admiring his agility and wait
until he gets cold and has to come in. His name is Lieutenant Philippe
Tailliez. His undersea gun works on the same theory as mine. Tailliez's
goggles are bigger than mine. He tells me where to get goggles and
fins and how to make a breathing pipe from a garden hose. We make a
date for a hunting party. This day is a big episode in my life."

From that day, Cousteau, Tailliez, and Didi were inseparable. They
called themselves *Les Mousquemers* (The Sea Musketeers), spending
every spare minute together in the water fiddling with JYC's camera,
figuring out new ways to dive deeper and stay longer, and frolicking
like children transported to an amusing new planet. Their watery
playground was not their natural habitat, and the three clever young
men were also thrilled with sorting out new ways to survive in it.
Except in the warmest months of summer, for instance, the Mediterranean Sea was chilly and simply staying in the water long enough to
hunt effectively was a challenge.

Without ever declaring himself the leader of their underwater
enterprises, Cousteau led his two willing friends into systematic
research on preserving heat. First, they dismissed the widely held
belief that coating one's body with grease would insulate it from the
cold. They found by trial, error, and talking with doctors at the navy
base that grease quickly washes away, leaving a film of oil that actually
increases the loss of heat. Cousteau's solution was to begin experi-

menting with a suit of vulcanized rubber, tailoring it like a set of overalls and patching it together with a heating iron. In the water he spent most of his time fighting its buoyancy and the irregular pockets of air that stood him on his head or flipped him on his back. He laughed at his comical failures and continued to believe that he or one of the others would figure out how to make it work.

Les Mousquemers also began studying the vagaries and effects of water pressure on divers. Air-breathing animals evolved in an atmosphere of oxygen, nitrogen, and a few trace gases, a sliver of air held in place around the earth by gravity. The actual weight of the molecules of air in a column extending from one square inch of any surface at sea level up to the edge of the atmosphere at about 120,000 feet is 14.7 pounds per square inch. Breathing is an unconscious mechanical process regulated by a pressure of precisely 14.7 pounds per square inch on a muscular diaphragm that contracts when the volume of the lungs decreases, reducing that pressure and triggering a demand for more air. Ninety percent of air molecules are in the first 7,000 feet of the atmosphere, though, and above that altitude the weight of the molecule, and therefore the pressure of the air, has dropped enough to make breathing difficult. Above 12,000 feet most people can't get enough oxygen and their bodies begin to fail.

Not only are air breathers unable to extract life-sustaining oxygen directly from water, but their bodies do not naturally adapt to the enormous increase in pressure caused by the weight of the water. While it takes roughly 120,000 feet of air to accumulate the weight of 14.7 pounds on one square inch of a surface at sea level, it takes only the weight of 33 feet of seawater on one square inch to equal 14.7 pounds, or one atmosphere. As a diver descends, any part of the body through which air circulates, including his lungs and sinuses, feels the effects of that air compressing under the pressure. Free divers rarely go below 60 feet, so the worst effect of 2 atmospheres is usually pain in their ears and sinuses. On the way down, he simply holds his nose and tries to breathe out; the air fills his ear canal, equalizing the pressure, and the pain disappears. A far more serious consequence of adding 2 atmospheres of pressure to the body of an air-breathing animal is an embolism. An embolism is a minute air bubble in the bloodstream that forms under pressure, travels to the brain or lungs, and expands when the diver surfaces, resulting in terrible pain or death.

Though *Les Mousquemers* were obsessed with experimenting with survival underwater, they never quit competing to see who could kill the most and the biggest fish. Didi usually came out on top. He would cruise on the surface breathing through his snorkel until he spotted a fish below, then execute what he called the *coup de reins,* which means "stroke of the loins." Cousteau called it a lightning dive. It consisted of bending from the waist, pointing the head and torso down, and snapping the legs into the air. The body of the diver forms an arrow aimed at the prey below, and with strokes of the arms and then the flippers, he can be 15 or 20 feet down in two or three seconds. Part of the trick was clearing the pressure in the ears and sinuses without slowing the attack. *Les Mousquemers* learned to do it with a kind of gulping yawn instead of by holding their noses.

Once, Didi bragged that he could spear two hundred and twenty pounds of fish in two hours. He made five dives and wrestled up four groupers and a palomata totaling 280 pounds before his time ran out. *Les Mousquemers* became famous for their bravado and skill, but they also angered traditional fishermen. Undersea hunting became such a fad on the Riviera after *Les Mousquemers* and other pioneers refined their gear and technique that the larger fish were disappearing from near-shore waters. The simple sling spears that could kill a fish 5 feet away evolved into spring guns and underwater harpoons fired with compressed air cartridges with ranges of 20 feet or more. Eventually, the governments in coastal villages listened to the complaints of the fishermen and banned the air guns.

For ages, humans had been the most harmless, helpless animals underwater. With their masks, snorkels, fins, and spearguns they became apex predators. Cousteau noticed that the big fish remaining near shore had already learned to hover just beyond the range of his speargun, seeming to know that the new predators were limited to short assaults on their territory. The deepest a man could go on a single breath of air while swimming free was about 130 feet; the longest he could stay down was about two and a half minutes. The balance between predator and prey was about to change.

3

BREATHING UNDERWATER

When testing devices in which one's life is at stake . . .
accidents induce zeal for improvement.

Jacques Cousteau

DESPITE THE GRIM CERTAINTY that war in Europe was imminent during the summer of 1939, one topic dominated the Cousteaus' lively dinner table in Sanary-sur-Mer—figuring out a way to breathe underwater. Cousteau's family now included a son, Jean-Michel, born in the spring of the previous year. Simone was pregnant with a second child, and she told Cousteau that whatever he did in the water better not kill him. Though she and her husband still celebrated their adventurous souls as their most primal connection, she had the instincts of a new mother for the practical realities of keeping her children safe. If the unthinkable happened and her husband died, she would have had no choice but to return to live with her mother and father in what for her would have been a domestic prison in Paris.

Much of the conversation, especially after one of Cousteau's many forays into the technical library at the navy base, centered on other underwater pioneers who had failed. Their experiments meant that *Les Mousquemers* did not have to repeat them. In 1825, an inventor named William James had combined a copper helmet, a waterproof tunic sealed at the wrists and waist, and an iron reservoir surrounding his torso from which he manually pumped air through a hose to his mouth. The reservoir carried enough air for a seven-minute, untethered dive, but swimming was out of the question because of the weight of the equipment. A diver was limited to clomping around on the bottom in less than 30 feet of water.

Over the next three decades, other inventors came up with more efficient ways to carry air underwater. One of these, Tailliez reported, was close enough to real free swimming that the French navy actually

Jean-Pierre, Jean-Michel, Jacques, Simone,
and Philippe
(PRIVATE COLLECTION)

had the gear in its inventory. In 1860, Benoît Rouquayrol and French navy lieutenant Auguste Denayrouze had patented a compact rig that a diver could carry on his back with no other special equipment. It consisted of a horizontal tank made of cast iron strong enough to carry a few minutes of air at low pressure, which could be refilled through a hose from the surface. The beauty of it was that the tank could be briefly uncoupled from the air hose, which gave the diver a taste of freedom beneath the sea.

After trying unsuccessfully to find one of the forty-year-old Rouquayrol-Denayrouze rigs, *Les Mousquemers* discovered that in 1870, novelist Jules Verne had equipped the divers of the submarine *Nautilus* with a fictionally enhanced version that allowed them to walk underwater for hours instead of just a few minutes. In 1878, an Englishman, Henry Fleuss, had come up with the idea of replacing oxygen from a reservoir through a manual valve and absorbing carbon dioxide with caustic soda, freeing a diver from any sort of connection to the surface. Fleuss fitted a standard metal diving helmet with double walls, and charged the space between them with oxygen at 16 atmospheres, or about 235 pounds per square inch. The diver wore a modified diving dress into which were sewn two bladders, front and back, containing pieces of sponge rubber soaked in a solution of soda lime. The diver wore a mask that covered his nose and mouth inside the helmet. He inhaled through his nose from inlet valves on both sides of the mask and exhaled through a flexible mouth tube. The pressure of exhalation pushed air laden with carbon dioxide through the two soda lime bladders, and finally out into the helmet, where it could be inhaled again.

The problem, everyone agreed, was deciding when to turn the control valve to release oxygen from the reservoir in the double wall of the helmet into the inside to enrich depleted air. A mistake could be fatal. Too much oxygen under pressure delivered to the lungs and blood of an air-breathing animal would cause convulsions and even death. Too little oxygen and the diver simply suffocated unless he surfaced immediately.

In 1934, a French navy officer, Yves Le Prieur, had combined a tank of compressed air instead of oxygen, a hand-controlled regulator, and a full face mask into a different kind of self-contained breathing apparatus. Advances in materials technology allowed foundries to cast stronger tanks to hold the air at higher pressures, which meant more time underwater. With the improved Le Prieur apparatus, an untethered diver could swim free for twenty minutes at 20 feet or fifteen minutes at 40 feet, manually releasing air from the tank on his chest whenever he needed a breath. But breathing underwater was far from perfected.

In August 1939, Leon Veche, a gunsmith aboard the cruiser *Suffern,* to which Cousteau was then assigned, showed up for dinner at the house in Sanary. Cousteau introduced him, telling the others that Veche had a fully equipped machine shop in which they were going to build a real self-contained breathing apparatus. Cousteau was convinced that it was only a matter of improving on the designs that had been around for seventy-five years.

A week later, Tailliez had requisitioned one of the Le Prieur rigs at the navy base, and *Les Mousquemers* and the rest of the household, including Simone, trooped to the beach to give it a try. Tailliez went first. Two minutes after submerging, he surfaced gasping and sputtering. The air flowed from the tank in powerful bursts when he opened the valve, he reported. Dumas and Cousteau each took a turn, with the same results. Because the valve was not calibrated to the depths at which the air was released, it was impossible to keep it from free-flowing and overwhelming a diver. In more tests, they got better at controlling the bursts of air and reached 50 feet, but all agreed that what they wanted was not brief dives to a single depth but longer dives at many depths. Le Prieur had given *Les Mousquemers* their first delicious sample of swimming free and breathing underwater, but they were far too busy wrestling with the air supply to hunt or run a movie camera.

Cousteau went back to Veche and his machine shop. They assembled a gas mask canister of soda lime, a small oxygen bottle with a bleed valve, and a length of motorcycle inner tube into a compact, self-contained system. With it, a diver would have to enrich his air supply with oxygen only every few minutes, which would give him plenty of time to get something done underwater.

In November, a month before Simone was due to give birth to their second child, Cousteau tested the device he called a rebreather. He left *Suffern* in an officer's gig with two sailors, motored a mile out of Toulon harbor near Porquerolles Island, jumped in the water, and submerged. For a few minutes Cousteau was in heaven. He exhaled, inhaled, opened his oxygen valve when the air tasted stale, and marveled at the view through the clear water of the offshore Mediterranean. Using the porpoise kick with his legs together that Tailliez had taught him, he imagined himself, finally, to be a creature of the sea. Visibility was 100 feet, the bottom 50 feet below. Cousteau gave himself another squirt of oxygen and instantly noticed the improvement in the air quality.

Breathing through the closed loop into the scrubbing canister, he could be as stealthy as a fish. Five minutes into his dive, he sneaked up on a school of several hundred chrome-bright giltheads with their distinctive red patches over their gills, getting to within 4 feet of the school before the fish spooked and disappeared into the distance. Cruising at about 30 feet, he saw a silver-blue bream hovering 15 feet below him. He circled around and in a maneuver similar to the wingovers he had learned in flight school, he dove down to see how close he could get to it. Cousteau was at 45 feet when, with no warning, his reverie was shattered by excruciating pain in his chest, back, and neck. His lips trembled uncontrollably. He lost his mouthpiece, gagged on a breath of salty water, and felt himself blacking out. In a final desperate move, Cousteau clawed at the buckle on his weight belt and released it. Seconds later, he bobbed to the surface a few feet from the boat, where his guardian sailors pulled him from the sea.

For a month afterward, Cousteau lived with sore muscles and Simone's indignation, while rebuilding his rebreather for another try. Cousteau incorrectly assumed that he had been poisoned by a buildup of carbon dioxide, so he refined that part of his rebreather. He went back to Porquerolles Island, this time with Dumas and Tailliez, and descended straight down to 45 feet to see if his changes to the CO_2 scrubber made any difference. This time, he convulsed so violently that he did not remember jettisoning his weight belt. He was limp, a dead weight, when Dumas and Tailliez hoisted him from the sea. The first thing he said when he regained consciousness was "It is the end of my interest in oxygen."

What Cousteau didn't know was that oxygen under pressure can be deadly to an animal that has evolved to breathe air containing precisely 20.947 percent oxygen at a sea-level pressure of one atmosphere, or 14.7 pounds per square inch. Because he brought his oxygen with him from the surface, his body had received more than one and a half times the amount it was designed to accept. Cousteau also had a particularly low tolerance for pure oxygen, and at the relatively shallow depth of 45 feet, his nervous system sounded the alarm that he was about to die by triggering the convulsions as the overrich gas moved through his lungs and into his bloodstream.

Stymied by the vagaries of breathing oxygen under pressure and the limitations of a hose to the surface, *Les Mousquemers* returned to breath holding. In September 1939, they were snatched from the pleasures of hunting and playing in the Mediterranean when one and a half million Germans crossed the Polish frontier. Two months later, Poland surrendered. The peacetime routine in Sanary-sur-Mer ended abruptly, as though a curtain had dropped. It seemed unlikely that the Germans would stop in Poland, so France prepared to defend itself. Cousteau was at Toulon or at sea on maneuvers every day, leaving Simone and the children alone in Sanary with instructions to begin stockpiling food. In April 1940, the Germans bombed airfields in northern France, and two months later took Paris. Pierre-Antoine Cousteau, who was stationed in northern France, became one of more than one hundred thousand prisoners of war.

After France surrendered, PAC spent a year in a hastily constructed camp near Sens, 140 miles east of Paris, enduring hunger, dysentery, and humiliation. Living outdoors in fenced cages, some of the men survived on food thrown over the fences by relatives, but thousands died. When PAC was released from the internment camp, he found his wife Fernande, their two children, and his mother alive but near starvation in a freezing apartment in Paris. His father, he learned, was waiting out the war at the Imperial Hotel in Torquay, England, still looking after eighty-seven-year-old Eugene Higgins.

With his prewar reputation as a fascist, anti-Semitic journalist, PAC found a job as the editor in chief of the pro-Nazi tabloid newspaper *Paris-Soir,* which had been started by Otto Abetz, German ambassador in Paris. Fernande remained with her husband in Paris, but he was able to arrange refuge in an alpine village near the Italian border for their children, who were then two and three years old. Pierre-Antoine Cousteau was soon part of the most powerful circle of French collaborators, which included the celebrated author Robert Brasillach, who became the editor of another pro-Nazi newspaper, *Je Suis Partout.* PAC attended a Nazi rally in Nuremberg, where he interviewed Adolf Hitler for an article in which he declared the Führer's terms for defeated France to be extremely generous. In 1942, PAC published his first book, the 120-page *L'Amerique juive* (*Jewish America*). Set in the context of the Cousteau family's two years in New York, the book's central thesis was that German Jews had brought anti-Semitism down upon themselves by methodically accumulating much of the nation's wealth and installing themselves in positions of power. He asserted that if left unchecked, German Jews would have completely taken over the European economy as they had in America, while at the same time grabbing political power by installing secret Jews such as Franklin Delano Roosevelt. In his book, PAC traced Roosevelt's lineage back to the Rossocampo family of

Sephardic Jews in Spain, which migrated to Holland, where they became known as Rosenvelt.

During the first year of the occupation, *Les Mousquemers* were scattered. Dumas became an army mule driver in the Alps. Tailliez was assigned to the destroyer *Valmy,* in charge of torpedoes. Cousteau was transferred from *Suffern* to the cruiser *Dupleix* as a munitions officer. A little less than half the country was under the control of a collaborationist puppet government led by the aging general Philippe Pétain, who ruled unoccupied France from his headquarters in Vichy. When France surrendered, all its territory, military forces, and resources passed into the hands of Germany in the occupied northern zone, and to the Vichy government in the south. Practically, however, the French fleet and bases on the Mediterranean remained in the hands of the officers and sailors of the defeated nation under only cursory supervision by the conquering Germans and Italians. During the first year after France surrendered, Cousteau supervised a crew of sappers clandestinely rigging explosives aboard most of the ninety ships in the harbor at Toulon, while continuing to live in Sanary. If necessary, France would scuttle its Mediterranean fleet to keep it from the Axis powers.

Though the Germans promised the Vichy government postwar status as a self-governing part of the Third Reich, French resistance forces in the unoccupied zone began forming immediately after the capture of Paris. Cousteau, like most of the officers in the Mediterranean fleet, held no illusions that the fascists were planning a bright future for their defeated enemies. As long as he was not dead or in a prisoner of war camp, he decided to fight the Germans and Italians in whatever way he could. Quietly, without telling his family or friends, he made sure his superiors knew that he and his cameras and his experience underwater were ready to serve the resistance.

4

SIXTY FEET DOWN

IN THE WINTER OF 1941, *Les Mousquemers* reunited in the relative peace of Sanary-sur-Mer in the unoccupied south of France. Cousteau and Tailliez were assigned to the detachments watching over the idle fleet at Toulon. Dumas just showed up one day, having fled south from the Alps on foot after his mule driving unit surrendered. During their off hours, *Les Mousquemers* hunted underwater, the fish they killed now vital to the survival of their families, who otherwise would have been living on tightly rationed bread, butter, and dried beans. The menu was grilled fish, baked fish, and fried fish, but everyone knew they were having a much easier time of it than their relatives in the occupied zone to the north. They also knew that things in the south were going to get worse. The beans, bread, and butter were growing scarcer by the day, and hunting fish was a solution to hunger only if the calories they provided exceeded the calories they consumed to kill the fish. A hunting free diver burns more calories than a stoker in a steel mill. As *Les Mousquemers* grew weaker from fatigue and their meager diet, they had to spend much more time in the water for each fish.

Cousteau's navy commanders knew about his obsession with diving and underwater photography and encouraged him to keep experimenting. They had no trouble imagining the military potential of free-swimming saboteurs and salvors. His diving research would also be a good cover story when undercover intelligence became necessary during the occupation by Germans and Italians that seemed sure to come. To informers and spies, who were everywhere around Toulon, Cousteau, Dumas, Tailliez, and the rest of the Sanary-sur-Mer diving fanatics looked very much like a bunch of friends playing with their toys in the ocean rather than spies and saboteurs.

Hunting took up most of Cousteau's free time, and he kept working on ways to breathe underwater that would consume less energy.

Skin diving, 1942. (Standing, left to right) *Tailliez,*
Cousteau, and Dumas (COURTESY OF WWW.PHILIPPE.TAILLIEZ.NET)

He had given up on oxygen rebreathers, but Maurice Fernez, who
had collaborated with Le Prieur on an early version of his rebreather,
had invented a lightweight surface feed system. It consisted of a flexible hose from an air pump on the deck of a tender to a free-flowing
valve that the diver held in his mouth. The mouthpiece vented extra
air into the water in a dense cloud of bubbles, but since the air hose
was flexible, and a Fernez diver wore only a mouthpiece and face
mask instead of a heavy helmet, the system let him swim free within
the range of the hose length. Cousteau liked it because it was simple
and had nothing to do with oxygen. Even though the constant cloud
of bubbles made filming and hunting impossible, he tried out the system in Toulon harbor.

Cousteau climbed down a ladder from a barge on which a gasoline-
powered air pump chugged, while Dumas and Tailliez tended the hose
line. He was wearing a new mask with a single oval glass plate sealed to
his face with pliant India rubber, a huge improvement over the aviator
goggles because it covered his face but still let him squeeze his nose to
compensate for the pressure as he descended. He also wore his swim

fins, and barely noticed the weight of the leash that tied him to the world above as he glided through the water 40 feet down. Cousteau was enjoying full breaths of air and flying above the worn, muddy bottom of the harbor when the bubbles stopped and he felt as though he had been hit in the chest by a giant hammer. The hose had snared on the gunwale of the barge, the roll of the sea had broken it, and Dumas could do nothing to warn him. The 5-atmosphere pressure of the air from the pump instantly dropped to one atmosphere. If Cousteau inhaled, his lungs could collapse as they struggled to equalize. He realized what had happened, stifled his urge to take even one more breath, and ascended before he drowned.

A few days later, Dumas was at 70 feet with the Fernez apparatus when the air line ruptured again. He had been hovering over the wreck of a freighter in the outer harbor when the bubbles stopped and he felt the pain of his lungs beginning to contract. Dumas stopped breathing immediately, but 70 feet was at the very edge of a diver's ability to free ascend with empty lungs. As Dumas clawed for the surface, his oxygen-starved brain started shutting itself down. He lost consciousness just as he broke water. Cousteau dove in, kept him afloat, and shook him back into the world. After that, *Les Mousquemers* gave up on "the pipe" as a way to hunt or shoot film underwater.

Even with the demands of his wartime navy duties and feeding his family, Cousteau's mind was never too far from his dream of making movies underwater. He imagined himself first as a filmmaker, then as a diver, knowing that he was in on the ground floor of marvels the world had never before witnessed. With the exception of music, which Cousteau enjoyed without reservation, he devoted all his energies to his ambition to make moving pictures underwater. As a teenager in Paris, he had seen the American remake of *20,000 Leagues Under the Sea,* produced in 1916 by Carl Laemmle from film shot by two brothers from Norfolk, Virginia. Now that he was trying everything he could think of to keep his camera dry and shoot movie film underwater, he read everything he could about the American brothers.

George and John Williamson were the sons of a clipper ship captain who killed time at sea tinkering with practical devices such as a collapsible baby carriage and an electric ship's signaling lamp. Captain

Williamson also invented absurdities such as a way to play golf on the ceiling using balloons. He finally left the sea to set up a ship fitting company in Newport News, where, in 1908, he built an underwater chamber for inspecting ships without sending down a diver. A riveted steel observation cylinder itself wasn't anything new, but Williamson had come up with a system of interlocking metal sleeves with canvas gussets fitted to a hole in the top of the chamber that extended up to the surface. The tube was 3 feet in diameter, big enough for a man to slide through while hanging on to rungs inside, and strong enough to withstand the pressure down to 60 feet. Captain Williamson worked on what he called his "hole in the sea" for a decade, but it never really caught on.

In 1913, two years after Cousteau was born, Williamson's sons, George and Ernie, fell under the spell of Thomas Edison's much more promising invention, the motion picture camera. They had also come across a magazine story about underwater artist Zahr Pritchard, who had built a bunker in the steep bank of his pond with a window through which he could observe beneath the surface. The next time the Williamson brothers were at home in Norfolk, they persuaded their father to drag out the hole in the sea, loaded it on a barge, and took it out into Hampton Roads at the mouth of the Elizabeth River. They didn't have a movie camera—almost no one did—but they crouched in their father's observation chamber with a still camera and took pictures of seaweed, pilings, and fish. With Ernie inside, George swam down to the window and held up a copy of *Scientific American* for a photograph. They sent the picture and their account of using the observation tube to the magazine, which published them the following month.

The Williamsons had no idea that they weren't the first men to take photographs underwater. In 1893, Louis Boutan, a French zoologist, had lowered a view camera sealed with wax and mounted in a 400-pound frame into the Mediterranean Sea off Banyuls-sur-Mer to take a ten-minute exposure of himself standing in a diving suit. Boutan took hundreds more photographs, experimented with magnesium powder lighting, and wrote *La photographie sous marin* to document his work. A few years later, American Simon Lake took photographs from inside his pioneer submarine, *Argonaut*.

For the Williamson brothers, though, a few snapshots were just the first step in their plan to make movies underwater. In 1913, armed with the photographs taken from their father's hole-in-the-water, they raised enough money to launch the Submarine Film Corporation to build and test a similar device for filming beneath the sea. Investors were charmed by their enthusiasm, the enormous publicity surrounding Edison's motion picture camera, and the mystery of the ocean. In a year the Williamsons had a new observation tube and chamber they named the Photosphere. They also bought a French-made Eclair camera, a 40-pound contraption of brass, iron, and steel with precision gears, a variable shutter, and a hand crank to roll film past the lens at sixteen to twenty-four frames per second. On February 21, 1914, Ernie and George loaded their camera, film, and the Photosphere aboard a steamer bound from Norfolk to the Bahamas.

Two months later, the Williamson brothers were on their way back to New York with 20,000 feet of exposed movie film. They had shot coral reefs, fish, staged scenes in which one of them in a diving suit walked around on the bottom discovering "treasure," and the bubbly plunges of local boys leaping from a dock. Their tour de force was a showdown between a diver and a shark, set up by weighing down the carcass of a horse to attract the sharks. The movie they eventually released from their first expedition was *Terrors of the Deep*, which critics hailed as "something never viewed before by mankind." The Williamson brothers were catapulted even further into moviemaking history when they joined Laemmle's Universal Pictures to produce *20,000 Leagues Under the Sea*. On October 9, 1916, the first scripted, eight-reel, underwater epic opened in Chicago to rave reviews. The Williamson brothers, Universal, and their investors made a small fortune.

"If the rest of the picture were discarded," wrote one critic, "the undersea scenes alone would be worth three times the price of admission."

Cousteau had been in awe of the Williamsons' work since seeing their film in Paris. Learning that they had made a fortune shooting movies underwater added to his inspiration.

In the spring of 1942, Cousteau found a 35 mm Kinamo camera in a Marseille junk shop, a ten-year-old relic with no lens. He bought it for $25. Leon Veche, the gunsmith who had become the fourth member of *Les Mousquemers,* built a waterproof metal housing for the big camera, fitting rubber seals around the winder, focusing lever, and trigger, and an optical-quality pane of glass through which to aim and shoot. A Hungarian refugee living in Sanary, Papa Heinic, who had been drawn to the energetic haven among Cousteau and his friends, ground a new lens for the Kinamo. The camera and housing was an ingenious contraption, the only one of its kind in the world. With it Cousteau could shoot the largest-size film available and produce the sharpest images of the world beneath the sea that had ever been seen.

The problem was that in Vichy France, all the 35 mm movie film was being requisitioned by the Germans for their gun cameras, airplane reconnaissance, and combat cinematographers. Cousteau scoured photography shops from Marseille to Nice with no luck, until he realized that he didn't need movie film. Any 35 mm film would work, and there was plenty of black-and-white Leica still film around. Cousteau bought every roll he could find. At home, he and Simone huddled under blankets, laughing like children in a nursery hideout as they spliced the thirty-six-frame strips into 50-foot reels that would give him three minutes of shooting time underwater. "I don't think anyone with common sense would do it," Cousteau said of his film manufacturing under the covers. "It was absolutely crazy."

"Absolutely crazy" became Cousteau's code for off-the-cuff inventions that worked. His enthusiasm for outcomes that only he could envision seduced everyone into helping him even when they had no idea how what they were doing fit into the grand plan. Leon Veche built the camera housing to withstand pressure down to 60 feet, about as far as a free diver can descend and still have time to focus, shoot, and surface on a single breath of air. He made the seals for the winder, focusing lever, and trigger from the design for a device known as a stuffing box, through which a boat's propeller shaft passed. Each seal on the camera was a hollow, threaded, male–to–female fitting that could be tightened to squeeze tarred jute around the extension shafts that controlled the winder and the focusing lever. If the fittings were too tight, the shaft and lever would not turn; too loose and water would pour into the housing, fouling the camera's clockwork machinery and possibly cracking Heinic's fine lens. Veche tested it with the camera replaced by half a brick and dummy controls.

After two weeks of diving with the brick, the housing was staying dry more often than not. *Les Mousquemers,* Simone, Jean-Michel, and Philippe trooped down to the sea to test it with the Kinamo inside. The camera and housing weighed 20 pounds out of the water. It was slung on a wooden shoulder brace about 3 feet long, which would also be used to line up a shot as though it were a speargun stock. Cousteau, Dumas, Tailliez, Veche, and Simone stood waist-deep in the water while Cousteau held the camera 2 feet down and squeezed the trigger. The first 35 mm underwater film ever shot captured a blurry image of the dark talus a few yards offshore and a pair of feet. When they opened the housing on the beach, the inside was dry. For the rest of the day, Cousteau dove deeper and deeper, triggering a few seconds of film at each level, working the controls to test the seals under pressure, and learning how to manage the bulk of the contraption underwater while holding his breath. Eventually, he reached about 60 feet. The housing held.

Three days later, one of the seals leaked and Cousteau surfaced with a housing full of water and a disaster inside. Back at the house, Veche plunged the camera into freshwater, explaining that the minerals in salt water would do much more damage and the rinse would help. For two days, the gears, ratchets, springs, levers, and the rest of the guts of the Kinamo were spread out on a plank and painstakingly cleaned. When Veche reassembled the camera, it whirred like nothing had happened. From then on, many evenings included dismantling, cleaning, and reassembling the camera, whose inner workings became as familiar to *Les Mousquemers* as those of their military sidearms. On other evenings, after successful dives, Cousteau developed the film and marveled at what they were doing.

Once Cousteau was confident that his equipment was more or less reliable, and that he could fix it if it failed, his theatrical instincts took over. He was ready to make an underwater movie. The story would not be one of the man–woman–villain farces that Cousteau enjoyed filming as a teenager. It would be a never-before-seen underwater adventure with a simple plot: A skin-diving hunter descends into the sea in mask, fins, and snorkel, armed with a speargun. In a wonderworld of fish and gorgeous submarine light, he glides gracefully through the water in search of prey. He fires, he misses. He fires again, he misses. He fires a third time, hits a big fish just behind the gills, and wrestles his catch to the surface. It was classic storytelling. Cousteau had never studied Henry James, but the great writer's formula for captivating an audience was instinctive to him: invent a hero, put him up in a tree, throw rocks at him, and get him down again.

Cousteau risked using the Fernez surface feed equipment again, but quickly confirmed his suspicions that the cloud of bubbles from the waste air made filming impossible. They scared off the fish that were essential to the story, and the hunter had to be able to kill one of them at its climax. There was no choice but to free-dive, which meant painfully slow progress. For six months, when the Kinamo was not lying in pieces on the repair table and *Les Mousquemers* weren't meat diving or on duty, they dove with the camera, shooting hundreds of feet of film in thirty-second takes.

Cousteau, Dumas, and Tailliez traded off as the heroes and cameramen. They filmed each other shooting and missing, shooting and hitting, corkscrewing through the water, and mugging as they swam straight at the camera. Sometimes, they shot background instead of action, scenes of coral, anemones, urchins, flatfish skittering along the bottom, and the colonies of sea life encrusting the giant boulders that had tumbled into the sea eons earlier from the coastal cliffs above the Mediterranean.

After a summer of diving, Cousteau borrowed a two-reel, hand-cranked editing console from the navy photo lab, and taught himself how to cut and splice film into scenes and sequences. In late October 1942, he finished his first underwater movie, *Par dix-huit mètres de fond* (*Sixty Feet Down*). Its public premiere was hosted by the German Internationaler Kultur Film before an audience of German officers and Vichy politicians at the Théâtre de Chaillot in occupied Paris. The showing and a reception afterward were arranged by his brother, Pierre-Antoine.

Soon after finishing *Sixty Feet Down,* Cousteau went to Marseille for briefing on a new assignment as a Vichy France naval attaché in Lisbon. On the night of November 27, Cousteau, Simone, and their sons were in a hotel near the waterfront when the roar of airplanes flying eastward woke them up. They went to the parlor of the hotel and tuned the radio to the free broadcast from Geneva. The announcer, fighting back sobs, said that Hitler had abrogated the treaty and was invading southern France with bombers, tanks, and five thousand German and Italian troops.

The Cousteaus returned to their beds but were shocked awake again at dawn, this time by the clatter of tanks and trucks roaring through the street below their windows. Twenty miles southeast, in Toulon, another armored column and a division of infantry were minutes from seizing control of the harbor. Admiral Richard Laborde gave the order his sailors had been preparing for but dreading for two years. In a dreadful cacophony of high explosives that seemed to last forever, the French navy scuttled its Mediterranean fleet. The battleships *Strasbourg, Dunkerque,* and *Provence* burned and sank under towers of black smoke, as did Cousteau's ship, *Dupleix,* seven other cruisers, seventeen destroyers, sixteen torpedo boats, six transport ships, tankers, minesweepers, and tugs. The Germans were able to seize only one destroyer, one torpedo boat, and five tankers. Toulon harbor was a sea of fire, fed by fuel gushing from the crippled ships, and the flames and smoke were visible from Marseille to Nice through the day and into the next night.

Jacques Cousteau was a sailor without a ship. The navy canceled his assignment and ordered him to stay in Toulon to film the carnage in the harbor. Simone and their sons went back to Sanary to prepare to flee to Paris, where Cousteau's mother, Elizabeth, his brother Pierre-Antoine, and his family had the benefits of PAC's privileges as the editor of a pro-occupation magazine. The Germans had garrisoned Toulon and surrounding towns with Italian troops, and Cousteau knew that once the occupiers felt they were on secure footing, things should eventually settle down on the Mediterranean just as they had in northern France. Conditions would be difficult, but he and his family would survive better at home, especially with the sea to feed them. If they were forced to flee, his plan was to return from occupied Paris in a month, maybe two.

During the few weeks it took for PAC to arrange his brother's family's escape from the chaos that had descended on the Mediterranean coast, Cousteau accepted an undercover assignment. The French resistance alone could not overthrow an occupation army, but it could gather information about troop strength and concentration. Liberation, if it came, would depend on the arrival of the Free French army from Algiers with their American and British allies. The more they knew about what they would find when they came ashore in the south of France, the better. Cousteau's contacts in the resistance were convinced that the Italians and the few Germans were so confused as they scrambled to gain control over the huge military and civilian populations of Toulon that they had no idea which of their own officers belonged in which offices. It presented an enormous onetime opportunity to gather vital information about their plans. Cousteau agreed. Wearing a stolen Italian uniform, carrying a Leica in a dispatch case, Cousteau simply strolled into the waterfront Italian headquarters and blended in with the bustling crowd of officers and men. During ten of the most dangerous minutes of his life, which would earn him France's highest military decoration, the Légion d'honneur, he photographed maps showing gun emplacements, wrecks in the harbor, and ammunition dumps. Then he walked away.

In Sanary, while packing to leave for Paris and not knowing if she and her family would ever return, Simone wrote a letter to her father to tell him they were coming. Retired Admiral Henri Melchior was a director of Air Liquide, living relatively comfortably under the occupation because the factories that produced compressed gas were industrial prizes for the Germans. They needed experienced people to run them. In a postscript to her note, Simone asked her father if anyone in his company might know something about building a demand regulator for dispensing gas. JYC, she explained, had abandoned the dangerous oxygen and hose systems for breathing underwater, but had lately become convinced that breathing compressed air from a tank was the answer if he could figure out a valve to regulate it.

5

SCUBA

I was playing when we invented the Aqua-Lung. I am
still playing.

Jacques Cousteau

LIKE MOST OTHER PARISIANS in the winter of 1942, Émile Gagnan
was scratching out a chilly existence, hoping that the Russians, British, and Americans would somehow manage to end the German occupation. He still had his job as an Air Liquide engineer in Paris,
though most of the oxygen, hoses, regulators, and the rest of the compressed air equipment flowing from the company's plants in France
were going to the Nazis. Gagnan was forty-two years old. The modest trajectory of his life had carried him from rural Burgundy, through
technical school in Paris, to the Air Liquide laboratory, where the
puzzles of liquefying, containing, and releasing gas under pressure had
held his interest for fifteen years.

Gagnan loved the clarity of physical laws, a world that was seen by
everyone but understood by only a select few. The transformations of
states of matter were particularly magic to him. A boiling teakettle on
a stove top offers a simple example of a liquid becoming a gas; rain, of
gas becoming liquid; and melting ice in a glass of lemonade, of the
transformation of a solid to a liquid. Under greater or lesser pressure,
the temperatures at which those transformations occur change. That is
why water in a teakettle on an ocean liner, Gagnan discovered, boils
faster than water in a teakettle over the same burner high in the Alps,
where the pressure of the air is lower than that at sea level.

If a gas is held in a confined space and subjected to pressure, it eventually turns into a liquid. Until the middle of the nineteenth century,
there were no containers strong enough to withstand the pressure of
compression, so it was impossible to liquefy gas except in theory. In
1873, Carl von Linde, a German engineer, had developed a practical

French Undersea Research Group scuba divers. (Left to right) *Cousteau,
Georges, Tailliez, Pinard, Dumas, and Morandière*
(COURTESY OF WWW.PHILIPPE.TAILLIEZ.NET)

way to convert ether into a cold liquid by compressing it with a pump
and using it to chill beer in Bavaria. Brewers already knew that beer
stored at lower than room temperature lasted longer and tasted better,
and slaughterhouses quickly caught on to the advantages of refrigeration over ice for storing meat.

Over the next ten years, von Linde improved his refrigerators to use
ammonia instead of the more expensive ether, sold hundreds of his
patented systems, and got rich. In 1894, he put his fortune to work
developing a process to liquefy ordinary air instead of ammonia. At the
time, that process had only limited practical applications, but it led him
to a next stroke of genius, the separation of oxygen from the other gases
in air while they are in their liquefied states. He removed the impurities
of water vapor and carbon dioxide from the air by drying it, then
pumped it up to 200 atmospheres, or 3,000 pounds per square inch, in
a metal chamber. Under pressure, the molecules of air squeezed
together, creating friction, which generated heat. He then passed the
air through radiators to remove the heat. The oxygen reacted to the
process of compression and expansion, repeated over and over, by
becoming liquid at very low temperatures.

Liquid oxygen is a lustrous, pale blue. It boils at precisely −183
degrees centigrade, at which point it can be distilled away from the

nitrogen, hydrogen, argon, neon, and other gases in ordinary air. Each separate liquid element turns back into a gas at a different temperature, so when liquid air is heated or cooled, each kind of gas can be drawn from the mixture and reliquefied to produce pure liquid nitrogen, hydrogen, argon, neon, or oxygen. As pressure increases, the boiling point decreases, so one of the keys to converting gas to liquid and back again was building containers that were strong enough to safely hold compressed gas at high pressure.

When pressurized liquid oxygen is released into one atmosphere, it instantly becomes a highly flammable gas with hundreds of industrial, military, and medical applications. It enables airplane pilots to breathe at high altitudes, and burners of all kinds to burn hotter. It is indispensable for removing impurities in the manufacturing of steel. One of the most important uses for high-tensile steel was in the creation of stronger and stronger pressure vessels, which in turn triggered a boom in the liquefaction of gas. By the turn of the century, companies around the world, including Air Liquide, were buying rights to the patents of von Linde, Wroblewski, Olszewski, and others. The transformation of states of matter built great fortunes on what was thought to be alchemy just a few decades earlier.

After Émile Gagnan finished technical school in 1927, he joined the ranks of thousands of other scientists and engineers who were parsing the intricacies of compressing gases, solids, and liquids and putting them to work. Air Liquide had expanded steadily since it was founded in 1902 by Georges Claude and Paul Delorme, who had licensed the patents for the processes of liquefaction and distillation of air. Claude and Delorme quickly realized that shipping a 220-pound steel cylinder containing only 6 cubic meters of compressed gas was a quick way to go bankrupt, so they designed a standard liquefaction and distillation plant to produce the gas closer to their customers. Their liquefaction plants sprouted overnight, and by the time Gagnan went to work for Air Liquide, the company dominated the market in Europe and Japan. Gagnan spent most of his time improving the hardware for producing liquid oxygen, nitrogen, hydrogen, argon, and neon, and the valves, gauges, tubing, and other equipment the customers needed to use the gases safely.

In occupied France, Air Liquide was among the most priceless spoils of war upon which the Germans were relying to increase their production of steel and supply oxygen to their pilots. At the laboratory in Paris, Gagnan and most of his colleagues did as little as possible to help their ancient enemy, but they brought their full energy to bear on designing equipment to improve life for the French. In December 1942, petroleum was in short supply, so Gagnan was working the bugs out of a regulator with which a farmer or a merchant could easily adapt the engine of his tractor, truck, or car to run on more plentiful methane or cooking gas. Simone Cousteau's father, Henri Melchior, a senior director of Air Liquide, knew about the development of the natural gas valve. When his daughter wrote to ask him to introduce her husband to an engineer familiar with demand regulators, Gagnan was the obvious choice.

A few days before the new year, after Paris had observed one of the most cheerless Christmases in its history, the navy officer and the engineer met in a workshop cluttered with valves, meters, tanks, and tubing in various stages of assembly. Gagnan, a quiet man with a distinctly formal demeanor, showed Cousteau to a separate inner office and motioned him to a plain wooden chair. He sat behind his desk, lit his pipe, and asked what he could do to help. For the better part of an hour, Cousteau took Gagnan through his experiences with rebreathers, surface-feed systems, and breath holding, pointing out the flaws in each. Finally, he asked the big question. Did Gagnan know of any way to simply carry ordinary compressed air in a tank that would flow into a diver's mouthpiece only when the diver took a breath? Gagnan turned to a shelf behind his desk, picked up a brown, rectangular box, and handed it to Cousteau.

"Maybe something like this," Gagnan said.

Cousteau examined the object in his hand. It was a hard, black, Bakelite casing about 7 inches by 5 inches by 3 inches, with a hollow tube about an inch and a half long and a quarter of an inch in diameter protruding from one side and a threaded metal fitting from another. Gagnan let Cousteau hold the thing for a long minute, then took it back and removed several screws to open the housing. He explained that the device was a demand regulator for reducing the pressure of compressed natural gas to feed it to an internal combustion engine in place of a gasoline carburetor. It dispensed a measured dose of gas: the rubber diaphragm over the exhaust tube closed in response to the drop in pressure inside the regulator, then opened again when the pressure on the other side equalized. One tube from the casing could be connected by a hose to the carburetor manifold, the threaded valve to the tank of natural gas. "Same kind of problems, you know," Gagnan said. "You have to reduce the pressure of the gas to feed it to an engine."

Cousteau and Gagnan were not the only inventors trying to find a way to safely breathe compressed air underwater. Their most threatening competitor was Georges Commeinhes, the son of the inventor of a breathing apparatus for firefighters who had a workshop in Saint-Maur on the outskirts of Paris, a dozen miles from the Air Liquide laboratory. Commeinhes built his fireman's rig with two tanks of compressed air at 2,000 pounds of pressure, a two-stage demand regulator, a single breathing hose, and a one-way exhaust valve on the mouthpiece. The French navy had been using it since 1935 and in 1939 had asked Commeinhes to adapt it for use underwater, but the war had slowed his research to a crawl. As a naval officer, Cousteau was familiar with the firefighting apparatus. He also knew that Commeinhes was working on a compressed air system for diving. Commeinhes had heard rumors from the Riviera that Cousteau was trying everything he could get his hands on to find a way to swim free and breathe underwater for long enough to work, hunt, and make movies. Both men understood that whoever was first to solve the puzzles and file the patents would be first to market with a dazzling invention that could be worth a fortune. Cousteau also knew that finding a way to breathe underwater and swim free would instantly make him an unstoppable force as a filmmaker.

On an unseasonably warm afternoon in January 1943, a half-dozen people gathered along an isolated back eddy of the Marne River east of Paris to watch Cousteau and Gagnan test their underwater breathing system. Simone was there, along with one of Gagnan's colleagues from Air Liquide and a storekeeper and his family from a nearby crossroads. Gagnan steadied the heavy backpack of three steel tanks as Cousteau waded from the low bank into the river. When he was knee deep, Cousteau bent from the waist and put his head underwater. Except for Gagnan's grip on his arm, the world above disappeared. He

crouched with his face in the water under the weight of the tanks, each
of them wrapped with an outer layer of wire for additional strength.
They had decided on three tanks because the point was eventually
to give Cousteau enough time to remain submerged for an hour at
60 feet. One or two of the standard industrial cylinders would not
carry enough air at the 110 atmospheres, or about 1,600 pounds of
pressure, at which it would remain a gas. The three tanks weighed
50 pounds on the surface but less than nothing underwater because of
the buoyancy of the gas. They had linked them through a manifold to
a slightly modified version of Gagnan's natural gas regulator, which
drew air equally from the tanks. Gagnan had designed a safety valve on
the manifold that stopped the air flow when there were only 300
pounds of gas pressure left in the tanks. When a diver could not draw a
breath because the reserve valve was closed, he could pull a metal rod
to release the remaining air, which would give him enough time to
reach the surface before running out completely.

Cousteau felt the weight of the tanks, sensed the rubber mouthpiece
clamped between his teeth, and felt the single hose leading from the
regulator rubbing on his shoulder. The water was murky, but he could
see the mud of the bottom, and he was breathing just fine. He nodded
vigorously, the signal that he was ready to submerge completely. Gagnan took his hand from Cousteau's arm. Cousteau stretched out on his
belly, pushed away from the bank, and sank beneath the shimmering
surface of the Marne. The shock of the bitterly cold water banished
any other sensations for the first few seconds, but then Cousteau heard
the strange sound of his own breathing. He heard a rasping rush of air
as he inhaled through the hose, the gurgling of a cloud of bubbles as he
exhaled through the exhaust valve in his mouthpiece, and the snap of
the diaphragm of the regulator as it responded to the changes in pressure as he breathed. It was working.

Cousteau glided away from the shore and let himself sink feetfirst to
the bottom, where his moment of elation dissolved in a cloud of bubbles as air flowed freely whether he inhaled or not. He could breathe,
but the bubbles from the exhaust valve on his mouthpiece obscured his
vision, as it had with the Fernez system. Free-flowing air was also too
wasteful for the system to be practical. Cousteau bounced his way into
deeper water, performed a "stroke of the loins" to put his head down
and his feet up, and most of the bubbles disappeared. But then he could

barely draw a breath. Experimenting with several other positions, he confirmed that the system let him breathe easily and did not free-flow only when he was perfectly horizontal. Like a test pilot, Cousteau reviewed his results to order them carefully in his mind. With his work done, he noticed again that the water was freezing cold. A few seconds later, he surfaced, waved to the anxious-looking people on the bank, and dog-paddled to shore, where Gagnan helped him from the river.

Gagnan could not swim, so he didn't test the system himself, but on the drive back to Paris, he relived the first dive of the self-contained breathing apparatus as Cousteau described it in detail. Gagnan rode in silence for a few minutes, envisioning a diver with his head up, and the mouthpiece above the level of the regulator. With the diver's head down, the mouthpiece would be below the level of the regulator. The problem must have had something to do with the difference in water pressure in different positions. Then, as though a light had clicked on, Gagnan had the answer. If a diver could inhale easily and the air did not free-flow when he was horizontal, all they had to do was relocate the exhaust port from the mouthpiece to the same level as the diaphragm, which would mean that the pressure on each would always be equal. It didn't matter if the diver was horizontal or not; it mattered only that the regulator and the exhaust valve were in the same plane. In Gagnan's workshop, he and Cousteau substituted a mouthpiece venting into a second hose running to an exhaust valve mounted inside a metal shield on top of the regulator. Gagnan considered the two-hose solution to be less than elegant, but it worked.

A week after the test dive in the Marne River, Cousteau's furlough ended. He had no choice but to return to Toulon with Simone and their sons. Cousteau, Gagnan, and Air Liquide knew that the self-contained underwater breathing apparatus was a breakthrough with enormous commercial potential, especially in sales to the navy. Scientists, too, might buy the equipment, which could revolutionize underwater research. Beyond that, the two inventors of scuba saw only limited applications for amateurs. Military and scientific sales would certainly be enough to justify an investment in production lines and a distribution network, so they began the yearlong process of filing for patents immediately. The apparatus they described in their application included the entire assembly of tanks, harness, regulator, double hoses, exhaust port, mouthpiece, and the reserve air valve. They called it

Scaphandre Autonome, or Aqua-Lung. Gagnan promised Cousteau that he would ship him a working prototype by late May or early June.

Back on the Mediterranean, the Cousteaus decided to live communally with Dumas; Tailliez; Tailliez's wife and newborn child; Claude Houlbreque, a former sailor on *Dupleix* who had been a cinematographer in civilian life; and Holbreque's wife. Together, they rented Villa Barry, a sprawling turn-of-the-century home with a vegetable patch and a garden of pines, in the seaside village of Bandol, two miles from Sanary-sur-Mer. Wartime shortages and the other hardships of the occupation were growing steadily worse on the Riviera, but sharing food-gathering chores, meals, and companionship at Villa Barry made life a little more bearable. Cousteau and Tailliez had orders to report for muster in the morning at the navy base and to keep their eyes open for unusual activity among the occupation troops. Otherwise, their days were their own.

In early June 1943, Gagnan charged the three tanks of the improved Aqua-Lung with compressed air, crated it up, and put it on a southbound freight train as part of an Air Liquide shipment. The crate, marked as scientific equipment, arrived in Toulon ten days later and was transferred to a local train for the brief trip to Bandol. When a messenger brought the word of its arrival to Villa Barry late on a warm afternoon, Cousteau drove to the station alone to avoid attention from the Italian troops in the rail yard. He was back home at dusk, and unloaded the crate into the workroom at the back of the villa. After a dinner of beans, bread, butter, and agonizing anticipation, the household gathered to look at the invention that Cousteau had told them was the stuff of wild dreams.

For weeks, they had talked about it whenever the topic of conversation turned to diving. Cousteau had explained the simple mechanism of the regulator, the strength of the new steel tanks that allowed the air inside to be compressed to many atmospheres, and the intricacies of the intake and exhaust valves that would allow a diver to breathe easily in any attitude underwater. Together, they speculated that hunting fish and lobsters was going to be as easy as plucking vegetables from a stall table in the market. Most of all, Cousteau insisted, the Aqua-Lung meant the end of experiments with dangerous gases and holding their breath to shoot film underwater.

Early the following morning, before the sunbathers were out, the household trooped in pairs through the pine garden to a quiet inlet with a gently sloping beach out of sight of the sentries in the city center. Dumas carried the Aqua-Lung, but when they reached the water, he helped Cousteau into its harness and followed his instructions for double-checking that the air was turned on, the tanks were secure, and the two hoses were firmly attached to the regulator. As the best free diver in France, Dumas would stay on shore to be ready if something went wrong. Simone, in mask, fins, and snorkel, would swim out to watch over her husband from above and signal to Dumas if he got into trouble. Cousteau spat into his mask and rinsed it in the sea, a trick *Les Mousquemers* had learned for keeping it clear of mist. He fitted the mask tightly to his face, covering his nose and eyes to his brow, clamped the mouthpiece between his teeth, looked around for a moment at his friends, and waddled into the water. When Cousteau was chest deep he stopped and lay facedown to gauge his buoyancy with the tanks of air on his back. He and Gagnan had designed the Aqua-Lung to be slightly buoyant in seawater because adding weight was simple and subtracting it was impossible. Dumas waded out and cinched a belt around Cousteau with 5 pounds of lead, but it wasn't enough. He added two pounds more, stepped back, and watched his friend sink slowly into the crystal clear water.

Cousteau breathed effortlessly, delighted by the distinctive whistle of air when he inhaled, the rippling of the bubbles over his head when he exhaled, and the snap of the regulator as it released each breath. He let his arms stream along his sides, fluttered his legs, and glided slowly over the sloping sand. The light danced down from the surface and flashed off the bottom until it gave way to a canyon full of dark green sea grass. Cousteau coasted to a stop. He exhaled until his lungs were nearly empty to find out what that did to his buoyancy. As expected, he sank slowly until he inhaled and began to rise toward the surface. Taking a single breath from his tanks turned him from a negatively buoyant object into a positive one. His lungs, he realized, were a sensitive ballast system. He steadied himself with his arms and swam smoothly down to about 30 feet. Cousteau felt a squeeze in his ears and sinuses, but no other effects of the pressure and no change that he could sense in the flow of air. The regulator was operating efficiently at 2 atmospheres of pressure.

Cousteau smiled into his mouthpiece as he reached the bottom of the little canyon, greeted by a flashing school of bream, round and flat as saucers. He hung on to one of the rough, limestone walls and did a quick check of his equipment, patting his harness and weight belt, shrugging his shoulders to be sure the tanks were riding well, and adjusting his mouthpiece. Cousteau looked up at the surface, which was shining like a rippled mirror. Directly above him, Simone was a small, silhouetted doll against the dazzling sheet of light. The doll waved at him. He waved back.

Cousteau held on to his rocky anchor and studied his bubbles on their way to the surface. They swelled and flattened into mushroom shapes identical to jellyfish as they rose through the water. Since the bubbles flowed from the regulator behind his head, the water in front of him was clear, which gave him a moment of elation as he thought about diving with his camera.

"I thought of the helmet diver arriving where I was on his ponderous boots and struggling to walk a few yards, obsessed with his umbilici and his head imprisoned in copper," Cousteau remembered about that moment. "On skin dives I had seen a helmet diver leaning dangerously forward to make a step, clamped in heavier pressure at the ankles than the head, a cripple in an alien land. From that day forward, we would swim across miles of country no man had known, free and level, with our flesh feeling what the fish scales know."

He looked again at the bream nosing curiously around him. They always return to the horizontal from a burst up or down, Cousteau concluded, because the horizontal must be the ideal attitude for moving in a medium eight hundred times more dense than air. Any other attitude required an expenditure of energy. Cousteau kicked and rolled through several revolutions on an axis from his head to his feet, turned a somersault, and did a barrel roll he remembered from flight school. He exhaled, sank headfirst to the bottom, balanced upside down on one finger, and laughed so hard he lost his mouthpiece. Taking a breath was slightly more difficult with his head straight down than in any other attitude, and Cousteau made a mental note to report that to Gagnan. He flipped upright, kicked hard, and soared upward through his own bubbles until he was just 10 feet below the surface. He swam out into deeper water and dove to 60 feet. Nothing he did changed the steady whistle, gurgle, and snap of his breathing. The regulator worked perfectly with his body in any attitude.

Three full tanks of air gave him sixty minutes at 60 feet. Cousteau had used up fifteen minutes. Despite the chill of the deeper water he was going to stay as long as he could. He swam over familiar limestone chasms that narrowed and turned into tunnels that had terrified him as a free diver afraid of being trapped inside with no air. Now he coasted fearlessly into one of them. The brilliant light from the surface dimmed as though it were being peeled away in layers, his tanks scraped against the rocks above him, and he felt the first twinge of claustrophobia. Cousteau's instinct for self-preservation overcame his passion to explore. He'd done enough on his first test dive. Before heading for the surface he rolled on his back to take a look at the roof of the tunnel, and saw that it was alive with lobsters. Hundreds of them were backed into niches in the limestone, their eyes glowing like fireflies in the dim light, their antennae flailing as they tried to get a fix on the giant intruder. Cousteau thought about his family and friends in ill-fed France, grabbed a pair of lobsters, backed out of the tunnel, and kicked for the surface. Simone saw him rising, swam down to him, and took their catch the rest of the way to the beach. He made five more trips into the lobster bonanza, Simone shuttled their catch to shore, and Jacques Cousteau became the first meat diver with the enormous advantage of being able to breathe underwater and swim like a fish.

A little over a month later, on July 30, Georges Commeinhes dove to 160 feet off Marseille with his firefighting apparatus modified for use underwater. Cousteau heard about the dive, but he didn't know whether Commeinhes had been able to solve the problems with the regulator that allowed him to swim free in any attitude with the Aqua-Lung.

6

SHIPWRECKS

DURING THE TWO WEEKS after Cousteau made his first dive, every-one at Villa Barry took a turn with the Aqua-Lung. Simone became the first woman scuba diver, and though the tanks were too heavy for the children to lift safely, they practiced breathing with their heads underwater in the shallows. The Aqua-Lung continued to work per-fectly, though the fear kindled by memories of sudden catastrophes with the Fernez pipe and rebreathers lingered. Each uneventful dive added to the suspicion that such astonishing freedom beneath the sea had to come at a higher price. Tailliez requisitioned a compressor from the navy base and with an unlimited supply of air they were making several dives a day. The divers reported on the ease of stalking prey, the delivery of air by the regulator at any attitude, and the utter bliss of swimming free underwater. After Tailliez's first dive, he led the household in a toast to escaping the world of the land and the abolition of gravity, after which everyone dug into heaping platters of fish and lobster.

"We were living in the middle of a war on pure fantasy and lots of beans," Tailliez wrote later. "When we got the Aqua-Lung, it was a miracle. We experienced in three-dimensional space the intoxication of diving without a cable. Back on shore, we danced for joy."

"Tailliez, Dumas, and I had come a long way together," Cousteau remembered about that time. "We had been eight years in the sea as goggle divers. Our new key to the hidden world promised wonders."

At the end of June, Cousteau sent word to Gagnan that their invention was working better than his wildest expectations. He asked him to file separate patents on the exhaust port and the air reserve valve immediately, and to send more Aqua-Lungs as soon as possible. With one Aqua-Lung, he could put meat on the table. But the

The poster for Cousteau's first underwater film
(© LAPI/ROGER-VIOLLET)

Kinamo underwater camera was in good shape, and he and Simone had stockpiled spliced reels of 35 mm film. Cousteau needed more than one scuba diver to make the movie he had been dreaming about for most of his life.

While they waited for Gagnan to ship them another Aqua-Lung, Cousteau and the other divers of Villa Barry concentrated on putting food on the table. "Tailliez went to the country and returned with five hundred pounds of dried beans, which we stored in the coal bin and ate for breakfast, lunch, and dinner, with an occasional maggot to break the monotony," Cousteau wrote later. He and the others cautiously stalked fish to supplement their tedious diet, being careful to avoid expending too much energy underwater.

Two more Aqua-Lungs finally arrived at the end of July. The timing was just right. In midsummer the Mediterranean was as warm as a bath, and the occupation troops had fallen into languor because nothing of consequence was being contested on the sultry southern coast of France. Cousteau continued his observations for the resistance, and started making occasional trips around Marseille, which he never explained to the rest of the household. On a visit in late July to gather information on mines and debris in the harbor, he came across a map on which the wreck of a British ship was pinpointed off Planier Island lighthouse.

The 5,000-ton steamer *Dalton* had left Marseille on a winter night in 1928 with a cargo of 1,500 tons of lead, sailed straight into the island, and piled up on the rocks. Lighthouse keepers rescued all hands—every one one of them drunk—and together they stood on the shore and watched *Dalton*'s stern settle into the sea, leaving the bow just below the surface to mark her tomb. She lay in clear water on a sloping bottom, unlike most sunken ships, which reposed in the murky, hard-used shallows of harbors, on surf-torn coasts, or in the unreachable darkness of the abyss. *Dalton* was perfect for filming a shipwreck.

Cousteau carried the identification card PAC had somehow procured that certified him as a marine biologist. "When I showed my *ordre de mission,* even the most brutal-looking Hitlerite was impressed," Cousteau remembered. "The word *kulture* (which was on the card) had a magic effect on them and we could work without much bother."

PAC's magic documents provided plenty of cover for the expedition to Planier Island with Tailliez, Dumas, cinematographer Claude Houlbreque, and Roger Gary, a friend from Marseille who knew the local waters. In early August, they took the weekly supply boat to the island with three Aqua-Lungs, nine spare tanks, the compressor, gasoline, spears, film, and the Kinamo. Leon Veche and Cousteau had modified the camera with a valve through which they could pressurize the housing from a tank of compressed air. Increasing the pressure to 10 atmospheres inside the housing exerted the same outward force as the inward force of water at a depth of 150 feet, as deep as they would go on any filming dive. Leaking seals were a plague of the past. Veche also built a new brace for the camera with a pair of pistol grips, one of which also held the shutter release. The Aqua-Lungs were identical to the prototype Cousteau used in the first test dives, with the same rectangular Bakelite regulators.

The lighthouse crew on Planier Island were frazzled from hunger and the anxiety of months of expecting an attack from the Germans or Italians. They welcomed the good-natured Frenchmen with their wild plan to dive beneath the sea to make a motion picture of their shipwreck. They were also delighted to share meals with their guests, who told them they would produce enough fish and lobsters to feed a bottomless pot of bouillabaisse.

For a free diver, swimming into a cave or the hull of a shipwreck was an invitation to disaster, but scuba divers with Aqua-Lungs could go anywhere. Still, when Cousteau, Dumas, and Tailliez swam into *Dalton's* gaping hatch 50 feet beneath the surface, they looked down the dark tunnel of the hold and knew that their freedom could be dangerous. The ship had broken into two pieces. They swam at a gentle downward angle through the maw of torn steel at the fracture until they could see the stern lying on the bottom like half a ghost ship with its two masts still standing. They had no way to know their precise depth, but estimated that they were at 100 feet, about 30 feet above the tempting wreckage of the stern. They were breathing easily, but with hand signals and head shakes they held a mimes' conversation in which they decided to surface instead of testing the limits of the Aqua-Lungs on that first dive.

Over their lunch of fish stew and bread, Cousteau, Tailliez, and Dumas reviewed their reconnaissance into *Dalton*. They were perfectly comfortable at 100 feet. The regulators clicked and gurgled with the same rhythm regardless of the depth. They felt fine after the dive, but they had talked to hard-hat salvage divers and knew that the pressure on the gases, fluids, and tissues in their bodies increased by one atmosphere for every 33 feet of depth. The risk of a painful or even fatal attack of the bends increased with every minute they spent at depths greater than 2 atmospheres.

That afternoon, they returned to the wreck with the camera. They wore enough weight so they began to sink as soon as they stopped kicking their fins. Cousteau hovered above to film Dumas and Tailliez as they descended along the wrinkled steel plates of the ship's corroding flank. Looking as comfortable as a pair of giant groupers, the divers cruised through the aquamarine water with their trails of bubbles glistening like moonstones and popping and burbling toward the surface.

Les Mousquemers had seen sunken ships before, but never the view that greeted them as they stood on the top of the rotting stern of

Dalton at 120 feet. They gazed down at the ship's twin propellers, deformed by their death throes when they turned their last revolutions in the sand. Then they checked each other's equipment, stepped off the stern, and settled the last 15 feet to the floor of the sea. Cousteau handed the camera to Dumas and swam around the ship with Tailliez. To sailors, shipwrecks are anathema, symbols of failure and bad luck that remind them of a fate that might overtake them at any moment on the sea. To Aqua-Lung divers, a shipwreck is a marvelous world alive with fish and mystery, through which they swim as if it were their natural home.

At 135 feet, most color had vanished from the spectrum. The hulk was a stark smudge of brown against the luminous haze filtering down from the surface. *Les Mousquemers* were deeper than they had ever been, deeper than all but a few hard-hat divers had ever been. With hand signals, they broke their reverie to let one another know that they were tiring more easily than they did in shallower water. As the camera rolled, Cousteau pointed to the surface, kicked away from the remains of *Dalton,* and swam up to the light.

Cousteau made his way underwater up the rocky slope along the side of the ship with no problems until he reached the bottom of the stone stairway leading into the sea from Planier lighthouse. At a depth of 10 feet, the scene beyond his mask blurred as his eyes refused to focus. Flashes of light blossomed like fireflies in front of him. Cousteau sat on the stairway until his sight returned, then left the water with no lingering effects. Dumas and Tailliez reported nothing like the symptoms Cousteau had experienced, which they assumed must be the result of congestion in his ears during decompression. *Les Mousquemers* had logged more than five hundred Aqua-Lung dives among them, but the incident reminded them that there were dangers in the depths that no one but they had ever encountered.

Air is a mixture of 78 percent nitrogen, 21 percent oxygen, and 1 percent trace gases including argon, carbon dioxide, methane, neon, helium, krypton, hydrogen, and xenon. To breathe, a body takes in air, consumes the oxygen, replaces some of it with carbon dioxide, but does nothing with all the nitrogen. At the surface pressure of one atmosphere, some nitrogen and oxygen dissolve in blood and tissues. As a diver descends, the pressure increases and more nitrogen and oxygen dissolve in your blood. Most of the oxygen gets consumed by a body's tissues, but the nitrogen remains dissolved.

Sixty years before *Les Mousquemers* were experimenting with the Aqua-Lung, another Frenchman, Paul Bert, discovered the effects of nitrogen under pressure on the bodies of air-breathing animals. Decompression sickness was known most commonly as caisson disease because bridge builders, tunnel diggers, and others who worked below sea level under pressure in sealed boxes called caissons were its earliest known victims. After working a ten-hour shift under pressure of more than 2 atmospheres, many men emerged from the caissons with crippling pains in their joints that sometimes proved fatal.

Bert theorized that the pain of caisson disease—some called it the bends, because a victim was often unable to stand upright or straighten his limbs because of the pain—was caused by some kind of imbalance in the relationships of the various gases in air. He tested his hypothesis by subjecting twenty-four dogs to pressures of 10 atmospheres, equivalent to a depth of 297 feet below the surface, and brought them back to one atmosphere in one to four minutes. Twenty-one of the dogs died; one showed no symptoms at all; the other two were in pain but recovered.

When Bert dissected the dogs afflicted with decompression sickness, he discovered that bubbles of nitrogen had formed in the tissues of their muscles and organs. In the dogs that had surfaced slowly, there were no bubbles. The nitrogen dissolved in body tissues as the pressure decreased gradually. When it decreased suddenly, nitrogen bubbles formed. The phenomenon was analogous to the absorption of carbon dioxide in champagne. With the cork in the bottle, the gas under pressure is dissolved in the liquid. When the cork is popped, the pressure is relieved and the carbon dioxide is released as tiny bubbles.

The solution was obvious. Caisson workers—and later divers—had only to return to the surface gradually to allow the bubbles of nitrogen time to pass harmlessly into the blood, or spend less time under pressure. Bert subjected another group of dogs to the same pressure as the first group, but brought them to the surface slowly in one to two hours. They suffered no ill effects at all.

Bert noted that once a caisson worker had been stricken with decompression sickness, the symptoms could be relieved by returning to the pressure at which the nitrogen had dissolved in his tissues. He tested this part of his hypothesis again with dogs, and discovered what would eventually become recompression treatment in hyperbaric pressure chambers for caisson workers and divers who had the bends.

Bert also discovered that breathing a gas that contained no nitrogen—pure oxygen, for example—would also reduce the symptoms of the bends. The gas containing no nitrogen simply forced the nitrogen bubbles out of the worker's tissues.

After Bert's pioneering work on decompression sickness, Scottish physiologist John Scott Haldane followed in 1905 with mathematical models showing precisely how much gas under pressure was absorbed by the different kinds of organs and tissues in an air-breathing body. Haldane used goats instead of dogs for his experiments because goats were closer in size to humans. From his research he derived tables that specified the amount of time a diver had to spend at various depths during his ascent to rid his tissues of nitrogen. Haldane's tables were published in 1908 in the *Journal of Hygiene,* adopted by navy divers in Europe and the United States, and two years later made available to the public. Cousteau had a copy.

Les Mousquemers dove on *Dalton* for two weeks, venturing inside the wreck to salvage crockery, silverware, glasses covered with coral, ship's lanterns, the oak steering wheel, and other loot. After a dive into what Cousteau reckoned was the captain's cabin, he returned to the surface with a crystal vial of clear liquid. Later, Simone uncapped the vial, took a whiff, and said the contents were a very fine prewar perfume.

Foreshadowing what would become generations of shipwreck divers who decorated their homes and garages with artifacts from their discoveries, Dumas's hunger for treasure was insatiable. "On *Dalton,* Didi gathered a lot of curious loot," Cousteau said. "He found stacks of crockery, silverware, glass bejeweled with corals, and a large crystal bowl. One day, he found a midden of ouzo bottles and thin Metaxas brandy bottles. He sawed off the oaken bridge wheel and dove repeatedly for dishes and silver. We suspected that he was collecting household gear for a wedding he had failed to mention."

For the rest of the summer, Cousteau and his band of divers filmed shipwrecks. During every minute underwater, they learned from experience. Near Marseille, the freighter *Tozeur* lay in 65 feet of water, the victim of a mistral that had blown it from its anchorage and onto a rock in the outer harbor. *Tozeur* taught them that the hull of a wreck could be coated with razor-edged clams that lacerated their bare flesh as they brushed against them. The cuts were painless underwater, but on the surface they hurt like bad insect stings. They named the clams "dog's teeth." *Tozeur* also introduced them to scorpion fish, which were ugly as toads and could inject a crippling venom through the spines on their backs. Tailliez was the first victim, after which everyone kept a wary eye out for the almost invisibly camouflaged nightmare.

Les Mousquemers tracked down wrecks with charts and firsthand accounts of sinking ships. In a cafe in the village of Cavalaire on the Côte d'Azur, a farmer told them the story of the Spanish freighter *Ramon Membru,* which had slammed into the rocks on an afternoon in 1925. The ship stuck firmly aground, a massive shape looming over the village, until a few days later, when a tug pulled it off and towed it to the harbor. That night, as *Ramon Membru* lay at anchor apparently safe and sound with its cargo of Spanish cigars, the ship caught fire, burned to the waterline, and sank. Cousteau, Dumas, and Tailliez found the remains of *Ramon Membru* a few hundred yards from the town jetty, all but invisible in the weeds that covered the burned and sea-torn hull. They filmed the ghostly outlines of the ship but found nothing of interest aboard the wreck. The highlight of the dives to *Ramon Membru,* and the film they shot there, was a herd of fish called *liches,* each the size of a man, that patrolled the wreck.

Off the coast near the village of Port-Clos they found a newly sunken fishing boat with nets and rigging still on its deck, which gave them the idea of filming a trawl net in action. Until then, fishermen had only been able to imagine what their nets were doing underwater. Cousteau hired a trawler captain and his boat, and set up with his camera on the bottom in 60 feet of water. The trawl slammed past Cousteau, destroying sea grass and bottom-dwelling creatures, while most of the fish leaped like rabbits to elude the gaping mouth of the net. The damage from a single pass of the trawl on the bottom devastated the near-shore areas of sea grass and other fish habitat. Dumas later hung, head down, on the tow rope to film what went into the net. The film from his camera revealed that a very small percentage of fish in front of the net were caught.

They explored and filmed another form of coastal destruction in Toulon harbor, where the 150-foot deep-sea tug *Polyphème* was scuttled along with the rest of the French Mediterranean fleet. *Polyphème*'s last assignment had been to open and close the antisubmarine net at the entrance to the harbor. A year after it was scuttled, the old tug lay in 60 feet of exceptionally clear water with the tip of its mainmast only 4 feet beneath the surface. *Polyphème* looked just like a tugboat under way with a slight starboard list, but sailing on the rocks and sand of the bottom instead of the surface of the sea. Inside, *Les Mousquemers* found that *Polyphème*'s crew had stripped their ship clean before opening the sea cocks and sending it to the bottom.

Nearby, according to a warning circle on the harbor chart, lay the wreck of the cargo ship *Ferrando*, which had gone down fifty years before. A local mariner told Cousteau that its location was marked by a buoy, probably placed there by a fisherman who had lost a net on the wreck. Dumas made the reconnaissance dive alone, descending 100 feet down the buoy line to explore *Ferrando*. The wreck had been plundered by hard-hat divers, who had cut a hole in its side that illuminated the cargo hold when he swam inside the hull. He found only a few china plates, and some lumps of coal that the minerals in seawater had turned from black to grayish green. The 300-foot-long wreck was festooned with nets and surrounded by the headstone-like upright black shells of giant mussels that sprouted from the sand on the bottom.

Exploring around the stern, Dumas found a single Japanese porcelain sake bowl, which he added to the crockery in his salvage bag. He checked his watch, which told him he had to ascend immediately or face a long decompression stage. He took a last look at *Ferrando*, turned to scan the open plain around him, and saw a rule-straight pathway cut into the sand and pebbles of the bottom. The strange road ran as far as he could see in the dimness of the light at 100 feet, a puzzling apparition that dominated Didi's report to Cousteau and Tailliez on the surface. They returned the next morning, but when they reached the approximate spot of the wreck the mooring buoy was gone. They dove all day but found no trace of *Ferrando* or the mysterious road on the bottom of the sea.

"Didi put the sake bowl and the crackled dish in his new house in Sanary, and a visitor who asks about them receives an interrogation on what he might know about Roman roads on the bottom of the sea," Cousteau said.

When they weren't exploring and filming shipwrecks, *Les Mousquemers* continued to test the limits of their Aqua-Lungs. In October 1943, after four months and hundreds of dives, Dumas talked Cousteau and

Tailliez into letting him be the guinea pig for a rigorously controlled test to find out how deep their wonderful new device could take them. They had been to depths of 135 feet several times since their descent to the stern section of *Dalton* off Planier light, suffering only the ear squeeze that had disoriented Cousteau. They carefully decompressed from dives during which they stayed for more than a few minutes on the bottom below 60 feet, and so far, none of them had experienced the symptoms of the bends.

On Maire Island off the coast near Marseille, Cousteau arranged for government witnesses from the local fishing village to verify the depth of Dumas's descent as measured by a 300-foot length of knotted rope. Dumas would simply tie his weight belt to the rope at his deepest point, and ascend to the surface with no fear of the bends because of his short time at the maximum depth. The signatures of witnesses would strengthen the case Cousteau, Émile Gagnan, and Air Liquide wanted to make to the French navy and other customers for the Aqua-Lung.

In the early evening, under threatening skies and with sea conditions at a whitecap chop and building, the test flotilla of two launches anchored off the island in 240 feet of water. They dropped the rope over the side, its end weighted by an anchor, and carefully tied the upper end to the rail of the boat. Cousteau descended first, stopping at the 100-foot knot from which point he would be able to reach Dumas quickly if he got into trouble. A minute later, Didi plunged past him, heavily weighted to speed his descent. Cousteau watched the bubbles from Dumas's regulator flowing away in the strong current, and saw Didi fighting to stay near the rope, which was streaming from the vertical. Dumas was flailing, and the bubbles from his regulator increased—a sure sign that he was hyperventilating in distress. Just as Cousteau was letting go of the rope to swim to the rescue, he saw Didi kick furiously. Seconds later, Dumas rocketed past Cousteau on his way to the surface.

Exhausted, Dumas told Cousteau and Tailliez what had happened. As he had passed the 120-foot knot, his vision had begun to blur and he started to obsess on the rope, the knots, its texture. Worrying about his eyes and the rope amused him rather than frightened him. He felt wonderful, counting knots as he went down, forgetting about Cousteau above him, the people in the launch, and the fact that he was diving to set a depth record. His ears were buzzing, he had a bitter taste in his mouth, and he felt so drowsy that he could barely keep his eyes open. He wanted to go to sleep, but had the vague feeling that he should stay awake. Dumas noticed the rope again, took off his weight belt, clipped it to a knot, and swam toward the light above.

In wartime France, *Les Mousquemers* knew nothing of the recent work of an American navy officer on the phenomenon of nitrogen narcosis, calling what had happened to Dumas "rapture of the deep." U.S. Navy captain A. R. Behnke studied the drunken euphoria that sometimes turned hard-hat divers giddy at depths of more than 100 feet and had killed many of them. He found out that nitrogen narcosis was caused by a combination of nitrogen saturation and excess carbon dioxide in nerve tissues, and could be alleviated by mixing helium into a diver's air supply.

When they pulled up the rope, Didi's weight belt was tied onto the knot at 210 feet. It was a world record for a free-swimming Aqua-Lung diver, to which the witnesses attested with their signatures on a certificate Cousteau had prepared in advance.

Back at home in Sanary-sur-Mer, the routine of the war years went on—scrounging for food, avoiding confrontations with occupying troops, and making the best of the worst of times. Cousteau and *Les Mousquemers* continued to shoot underwater and test the limits of the Aqua-Lung. They edited hundreds of feet of film into a movie called *Épaves* (*Shipwrecks*). It was under thirty minutes long, with no story line, but the moving picture of men swimming underwater around sunken ships enchanted everyone who saw it.

Among the first audiences to see *Épaves* was a roomful of admirals in Toulon who were stunned by the obvious military potential of the Aqua-Lung. Georges Commeinhes, Cousteau's closest competitor in the race to find a way to breathe safely underwater, was killed just days before the end of the war. During the liberation of Strasbourg, he was in command of a tank destroyed by a satchel charge thrown by a retreating German. Cousteau, Gagnan, and Air Liquide were then unopposed in their campaign to sell their Aqua-Lung to the French navy.

7

THE FOUNTAIN

ÉPAVES WAS AN INSPIRED promotional tool, an example of Cousteau's unique talent for forging ahead on instinct instead of detailed planning. He made the movie because he wanted to demonstrate that film of the exploration of a shipwreck could be thrilling and entertaining. That it also proved to be a captivating advertisement for the Aqua-Lung seemed like an unintended consequence, but that really was part of Cousteau's goal all along, even if he could not articulate it. The admirals in charge of the French Mediterranean fleet placed an order for ten Aqua-Lungs, which was a financial victory for Cousteau and Air Liquide. More important, they recognized that Cousteau himself was more valuable to them than any single piece of equipment and made sure that he had what he needed to keep inventing new ways to explore and work underwater.

Soon after victory in Europe in the spring of 1945, the French navy created the Undersea Research Group (Groupe de Recherches Sous-Marine). Tailliez, as the most senior officer, was its commander. Cousteau was the deputy commander. Tailliez and Cousteau made their first decision together to hire Dumas as a civilian adviser and chief diver. Three petty officers, Maurice Fargues, Jean Pinard, and Guy Morandière, completed the group, becoming Aqua-Lung divers after Dumas ran them through a crash course. Left largely to its own devices, the Undersea Research Group was housed in a single office on the ground floor of a warehouse on a pier in Toulon, then went to work scrounging supplies and material from navy surplus dumps. Tailliez commandeered a small launch, then a 78-foot landing craft. Cousteau talked the motor pool out of two trucks and a motorcycle, telling the officer in charge that they were for a newly created division of the powerful National Marine Institute.

Cousteau's family in Sanary after the war. (Left to right) *Simone, Philippe, Jacques, nephew Jean-Pierre, Jean-Michel, sister-in-law Fernande, niece Françoise, and Cousteau's mother, Elizabeth* (PRIVATE COLLECTION)

Though Cousteau, Tailliez, and Dumas would much rather have been training divers to thoroughly explore the Mediterranean with Aqua-Lungs, the first assignment of the group was clearing French harbors that were littered with mines, shipwrecks, and unexploded munitions. For a year, they investigated wrecks and cleared the sea lanes. It quickly became obvious that a diver with an Aqua-Lung could accomplish far more than a hard-hat diver in a given time, simply because he could find the mines faster. They invented an underwater sled that could carry a diver and be towed at 6 knots behind their boat. With it, they cleared the harbor at Sète of mines in a little over a month, a job that would have taken conventional divers four or five times longer.

When demolitions work tapered off, Cousteau led an inland diving expedition to the village of Vaucluse near Avignon, where a legendary spring emerges from the base of a 600-foot limestone cliff. The Fontaine-de-Vaucluse, memorialized in the fourteenth century by the poet Petrarch and more recently by the Bard of Provence, Frédéric Mistral, was one of the world's great hydrological mysteries. No one had been able to explain why a calm watery cavern turned into a gushing torrent spewing millions of gallons of water into the Sorgue River for five weeks every spring. Hard-hat divers had tried and failed to find the source of the celebrated fountain, which one puzzled scientist called "the most exasperating enigma of subterranean hydraulics." Cousteau knew that good publicity about divers using the revolutionary Aqua-Lung made the acquisition of equipment, boats, and supplies from the navy much easier. If he, Tailliez, and Dumas could solve the thousands-year-old mystery of the Fontaine-de-Vaucluse, their discovery would make sensational news and the navy would become even more supportive of the Undersea Research Group.

Before asking the navy for permission to go to Vaucluse, *Les Mousquemers* carefully evaluated the risk of a dive into a dark, underground river. Tailliez drew up a list of seven specific dangers:

> There was the instinctive repugnance to diving underground. There was the cold, for the water in the spring was no more than twelve degrees centigrade. There was the darkness, which our flashlights could pierce but feebly—if they did not fail altogether. The rope might part, leaving us to extricate ourselves from an uncharted maze. There might be a fall of rock. There might be suction or underground currents might pin us in some corner. And, finally, there was the danger of intoxication of great depths.

"I despise danger," Cousteau said. "I am not a thrill seeker but an explorer who intends to return myself and my men safely from every dive I make."

Before dawn on August 27, they left Toulon in one of their trucks carrying the new rubber diving suits with which they had been experimenting, four Aqua-Lungs loaded the night before with air from a new compressor, masks, fins, cameras, and coils of mountaineering rope. Using the reports from the earlier hard-hat attempts, which had reached a depth of 120 feet, they planned their dive as though they would be climbing a mountain instead of descending down a sloping tunnel filled with water.

A crowd of villagers, caving experts, and Simone Cousteau looked on from the rocky lip of the crater. Simone appeared none too happy about the dive into a dark cave, standing with her arms folded across her chest glaring at her husband as he lowered the weighted end of a 400-foot guide rope from a canoe. The weight stopped at the 50-foot mark. One of the group's newly trained petty officer divers plunged into the water wearing an Aqua-Lung but not a diving suit and freed the weight, which had snagged on a triangular rock that almost completely blocked the tunnel. He returned to the surface shivering uncontrollably. The guide rope had stuck again at 90 feet.

Cousteau and Dumas, dressed in heavy woolen underwear, squirmed into their new diving suits. Even during the warmest months, Les Mousquemers returned from long dives at depths over 60 feet in the early stages of hypothermia. They tried coating themselves with grease, which turned out to be worse than nothing at all. Most of the grease quickly washed away, leaving a thin coating of oil that increased the loss of body heat. If they could have injected the grease under their skin it would have worked fine, but otherwise, they needed a second skin. They sewed sheets of vulcanized rubber into a full-body suit, but discovered that the air trapped inside it produced uncontrollable buoyancy as the pressure changed during a descent or ascent. They spent most of their time underwater fighting the buoyancy or hanging upside down when all the air rushed to the legs of their suits. Just before the dive into the Fontaine-de-Vaucluse, they figured out how to maintain a constant volume of air inside their suits with escape valves at their neck, wrists, and ankles. They could replace lost air with exhalation of breath under the edges of a mask that vented into the suit.

Cousteau and Dumas roped up like mountain climbers, with a 30-foot length between them, and checked each other's air valves and equipment. Standing on the lip of the crater, they were loaded down like pack animals, each with three air cylinders, foot fins, dagger, and two large waterproof flashlights. Dumas wore a red-colored face mask, Cousteau blue, so their surface team could quickly identify each man. Cousteau carried 300 feet of line, coiled in three pieces. When they reached the end of the weighted guide rope, he would pay out his coils as they entered what they assumed would be another chamber of the cave. Dumas carried a small cylinder and regulator as an emergency air supply, and an alpinist's ice ax.

They went over the code they had worked out for signaling Maurice Fargues, who was in charge of tending the rope on the surface. One tug on the rope meant tighten the rope to clear a snag. Three tugs meant pay out more line. Six tugs meant pull us up as quickly as possible. Underwater, Cousteau wore a mouthpiece he had invented through which he could shout brief commands to Dumas, who wore a regular mouthpiece but could only respond with nods of his head and hand signals.

Cousteau and Dumas struggled into the water under their heavy equipment, and felt the now familiar relief as buoyancy made them weightless. They bobbed on the surface for a minute, made final checks of their regulator valves, and eyed the crowd on the lip of the crater above them, which had swollen to more than a hundred curious people. In the front rank stood a young, black-clad priest, whom Dumas and Cousteau assumed had arrived to oversee their departures if the worst happened.

The key to understanding the annual gusher from the Fontaine-de-Vaucluse, hydrologists had told them, was finding an inner chamber of air in which pressure could build up to discharge the gusher explosively in the spring. The goal of the first descent was to find the point at which the tunnel hit bottom and began rising to an inner air chamber. Cousteau and Dumas dropped into a narrowing tunnel about 16 feet in diameter, traveling along a rock face sloping downward at a 30-degree angle. Fifty feet down, they found the boulder on which the guide rope had snagged, and slipped past it through an opening barely big enough to accommodate a man wearing an Aqua-Lung.

Cousteau and Dumas had not imagined the frightening darkness into which they descended as they passed the boulder. The faint green blip of the entrance to the tunnel disappeared entirely. The water contained no diatoms or plankton, which in the ocean reflect the beams of flashlights; here in the cave their lights illuminated only coin-size patches of the wall down which they crawled. Cousteau glanced above him and saw that Dumas was braking his own descent with his feet to maintain his distance on the rope, but in doing so he was kicking big chunks of limestone downward.

At 90 feet, Cousteau breathed easier when he found the pig-iron weight of the guide rope resting on a ledge, right where it was supposed to be. He had learned he was more susceptible to the rapture of the deep than heavier, less leanly muscled men, and instantly recognized the beginning of its fuzzy embrace. Cousteau fought through the narcosis, remembered that he was supposed to do something with the pig-iron weight, and kicked it off the ledge with his heel. He did not know he had lost the coils of ropes on his arm, didn't know that he had failed to tug the line three times to ask for slack to allow the guide rope to sink deeper, and had forgotten even that Dumas was 30 feet up, which would have explained the irritating rocks pelting him from above.

Cousteau had a blinding headache, but he continued to descend. A minute later, he landed standing up on what seemed to be the floor of the cave. Rocks, dirt, and some debris that looked man-made surrounded his feet. He checked his depth. One hundred and fifty feet, but the gauge was full of water. That had to be wrong, Cousteau thought. They were at least 200 feet beneath the surface, and 300 feet from the mouth of the slanting tunnel. Cousteau followed his bubbles streaming upward but not into the shaft through which they had just descended. He was apparently at an elbow in the tunnel. Still in the tunnel, Dumas struggled with his suit, which had ruptured and was filling with water. He looked like a partially inflated balloon.

Cousteau's rapture suddenly filled him with the urgency to explore the upward shaft, which might lead to the solution of the mystery of the fountain. He shouted through his vocalizer mouthpiece, telling Dumas to stay at the rope while he swam away and up to look for the shaft. Dumas was woozy, deep into narcosis. He thought Cousteau was shouting at him because he needed air from the emergency Aqua-Lung, and lunged down into the darkness after him. Now both divers had left the guide rope, their only hope for getting to the surface because they could no longer follow their bubbles to the surface in the terrifying darkness of the cave.

Cousteau snapped back to sanity for a moment. He saw the faint light from a flashlight, swam toward it, and crashed into Dumas, who was limp and barely conscious. Cousteau looked into Didi's mask and saw his eyes rolled back into his head. Dumas woke, seized Cousteau by the wrist, and pulled him into a bear hug. Cousteau twisted free, and frantically swept the beam of his flashlight over the floor of the cavern. There was no current, so they had remained near the pig-iron anchor of the guide rope, and there it was. Cousteau reached into the darkness, grabbed Dumas, and saw with horror that Didi's jaw was slack. His mouthpiece had slipped from his mouth. Cousteau jammed the mouthpiece back in, grabbed the rope that still tied the two of them together, and started clawing his way upward, towing Dumas in his heavy, waterlogged suit behind him. He started climbing up the rope hand over hand, but on the surface, Fargues interpreted the first three tugs as a call for more slack on the guide rope. To Cousteau's horror, he felt the rope fall. When it stopped falling, he tried to climb again. More slack came down. Cousteau scrambled up the steep slope of the tunnel wall, thinking it was his only hope. Only then did he remember that six tugs on the rope meant pull everything up. Cousteau tugged six times, felt tension on the rope, then slack. Its 400-foot length was snagged by friction on the tunnel walls. The emergency signal was not getting through to Fargues. He looked down at Dumas hanging like a bloated sack beneath him, and pulled out his knife to cut himself free of his best friend, who was certainly dead. As his knife touched the rope tying him to Didi, Cousteau was yanked firmly upward by the guide rope. He put his dagger away and hung on for the sixty seconds it took to return to the world of light.

Five minutes into the dive, when Cousteau and Dumas had reached the elbow in the tunnel, Simone Cousteau had seen the bubbles stop gurgling to the surface. She could not stand to watch the still water, and fled to the village, where she ducked into a cafe and ordered a brandy. Soon after she sat down, a man running from the direction of the fountain was crying out that one of the divers had drowned. Simone grabbed the man as he passed. What color was the mask of the dead man? Red, the man replied. Cousteau's mask was blue. Simone felt the weight of mortal dread lift but only for a moment. The man she loved second only to her husband was dead. Dazed, she staggered back up the path to the base of the cliff to face the horrible truth. There she received one of the most wonderful gifts of her life. Both Cousteau and Dumas stood warming themselves over a barrel of burning gasoline, gesturing wearily to Tailliez and the others.

Later that day, Tailliez and Guy Morandière, who had been with the group since the beginning, descended into the fountain wearing only long underwear and lightweight belts so they would remain positively buoyant. At 120 feet, both divers felt the unmistakable onset of nitrogen narcosis and aborted their descent. After Tailliez signaled with six tugs, Morandière watched in horror as his partner whipped out his dagger and started slashing at the rope. Morandière swam under Tailliez, grabbed his ankles, and kicked for the dim green opening above them. On the surface, the crowd gasped when they saw Tailliez break water surrounded by a cloud of blood. In his rapture, imagining that he was entangled in the guide rope, Tailliez had slashed at it and in the process cut the fingers of his hand to the bone.

The disasters in the Fontaine-de-Vaucluse should not have happened, the divers concluded on the drive back to Toulon. They were underwater nowhere near long enough to have been stricken with nitrogen narcosis. They wondered if somehow the clear, still water in the cavern had created different pressures on their bodies than the sea. Dumas suggested that perhaps they weren't narced at all, but rather had suffered from some unknown reaction to fear in the absolute darkness. Or maybe there was something wrong with the air they were breathing. The following day, they analyzed the remaining air in their cylinders, and discovered that it was contaminated with six times the amount of carbon monoxide in normal air. The carbon monoxide could have come from only one source: their new compressor. They fired it up, attached a tank for refilling, and saw that the air intake of the compressor was sucking in exhaust from the gasoline engine. Under the pressure of 5 atmospheres, the carbon monoxide would have killed all of them in twenty minutes.

A month after they failed to uncover the secret of the Fontaine-de-Vaucluse, *Les Mousquemers* realized their far greater ambition to establish themselves as underwater cinematographers. In 1939, the French minister for arts and education had proposed the creation of an international event to celebrate motion pictures, naming the Mediterranean seaside resort town of Cannes as its site. This First International Film Festival, which would have been presided over by Louis Lumière, had been postponed until the autumn of 1946. It opened on September 20 for a two-week run, becoming the first major cultural event in postwar Europe.

Producers from a dozen countries presented twenty-three films, including Billy Wilder's *The Lost Weekend* from the United States; *Brief Encounter,* by David Lean, from Britain; *The Prize,* by Alf Sjoberg, from Sweden; and *Épaves,* by Jacques-Yves Cousteau, Frédéric Dumas, and Philippe Tailliez. The festival was more a film forum than a competition, because every film entered won a medal. *Les Mousquemers,* their families, and friends savored the gasps of the audience in the darkened theater watching men swim like fish as they explored shipwrecks that had never before been seen by human eyes. At the end of the festival, in the great hall of the Casino de Cannes, *Épaves* was awarded the special prize from the Center for the Arts, Literature, and the Cinema. Tailliez stood on the dais during the presentation looking like a slightly awkward sailor on shore leave, Dumas like a wild animal indoors for the first time. Cousteau, however, was beaming. He felt very much at home with the applause of the crowd and having all eyes focused on him but was far more thrilled by the possibility that he might really be able to make a living as an underwater cinematographer.

6

MENFISH

A WEEK AFTER Cousteau savored his triumph at Cannes, word reached him from Paris that his brother, Pierre-Antoine, had been arrested as a Nazi collaborator. During the war, Cousteau and PAC had chosen different sides, but they had never renounced each other. Their family bond meant more to them than national loyalty or the dismal business of doing whatever was necessary to survive and protect their wives and children. As a member of the resistance, Cousteau may not have agreed with PAC's politics or his loyalties during the war, but he could not forget the beloved PAC who had made his early childhood bearable.

"He was my brother," Cousteau said twenty years later. "Nothing else mattered."

As the American army massed at Chartres for the assault on Paris on August 17, 1944, PAC and his wife had rushed to the rue des Pyramides to join a truck convoy the Germans had organized to evacuate their most highly cooperative French collaborators. Already, broadcasts from radio stations in England included reports of in absentia death sentences for the most notorious French turncoats. The scene on the rue des Pyramides was chaotic amid rumors that resistance partisans were also assembling nearby to execute them en masse when the Allied army took the city. More than twenty thousand collaborators fled Paris that night. The Cousteaus set off with a small band determined to find refuge in Italy. At the Austrian-Italian border they changed course, fleeing into the Alps and eventually surrendering to American troops as a much better alternative than the Free French, who would have shot them on sight.

In the Allied prison camp at Landeck, PAC posed as a Pole to hide from visiting French officers, who had come to claim their traitors. He persuaded the American commanding officer of the camp to release

Fernande to live free in a nearby village, and was even granted permission to visit her after he swore on his honor to return and never try to escape. A few weeks later, Jacques-Yves arrived unannounced with false passports and transit visas that would get his brother and his family first to Spain and then to South America. PAC refused to leave, telling Jacques that he had promised not to attempt an escape. The brothers argued. Jacques told Pierre-Antoine that his honor was already compromised as a collaborator, but family loyalty meant more than honor, acts of desperation during wartime, or political enchantment. PAC refused. Cousteau was incredulous. How could his brother not think of his own family before all else? PAC and his promise to a prison guard would be worthless to Fernande and his children when he stood against a wall smoking his last cigarette.

The French army eventually identified Pierre-Antoine Cousteau, seized him from the Americans, and transferred him to a military prison outside Paris. PAC knew that his chances were not good. Robert Brasillach had been executed despite a public outcry against killing so revered an author, regardless of his wartime crimes. There would be no dramatic outpouring of support for a relatively obscure newspaper editor. The volleys of firing squads echoed dozens of times a day in France's prison towns, sending men far more luminous than he into eternity. PAC's trial for treason was brief, little more than a recitation of undeniable facts about his very public life during the German occupation.

Jacques Cousteau testified in his navy uniform wearing the crimson ribbon of the Légion d'honneur awarded to him for his undercover work with the resistance. Cousteau knew he was there to plead not for acquittal but against the death sentence. PAC's guilt as a collaborator was undeniable.

On November 23, 1946, the tribunal ignored Cousteau's plea and sentenced Pierre-Antoine Cousteau to death by firing squad, setting April 6, 1947, as the date for his execution. A month later, as part of a general amnesty declared to heal the tormented nation, his death sentence was commuted to life imprisonment at hard labor.

During the first months after the war, the tribulations of Jacques Cousteau's family in Torquay, Paris, and the military penitentiary at Clairvaux produced a steady hum of anxiety. He refused to let it dis-

(Left to right) *Jacques-Yves, Daniel, and*
Pierre-Antoine Cousteau during World War II
(PRIVATE COLLECTION)

tract him, however, from the revolution he was leading with the
Aqua-Lung. Simone and their sons Jean-Michel and Philippe were
safe with him in Sanary-sur-Mer. His parents, Daniel and Elizabeth,
had taken PAC's wife and children to England so they would not suffer the pains of being the family of a convicted traitor. There was simply nothing Cousteau could do about his brother. As always, his
primitive sense for knowing what he could change and what he could
not allowed him to live vigorously in the present.

Meanwhile, Émile Gagnan shipped a steady stream of Aqua-Lungs
to Cousteau at the Undersea Research Group base in Toulon for training submariners, munitions experts, spies, and reconnaissance teams to
scuba dive. Gagnan had replaced the original rectangular Bakelite case
of the regulator with a round metal housing with two cast-metal horns
to which were attached specially made flexible rubber hoses leading to
a single mouthpiece. Each regulator bore a plate with the engraved
inscription *Scaphandre Autonome Cousteau Gagnan*. There had been no

publicity about the Aqua-Lung, except for a slight flurry of interest after *Épaves* won the prize at Cannes. Cousteau and Gagnan knew, however, that reports of dives to greater and greater depths and sensational discoveries beneath the sea would eventually find their way into newspapers, magazines, and newsreels.

The first internationally covered story on the Aqua-Lung was one that Cousteau, Gagnan, and the Undersea Research Group would much rather have done without. On September 17, 1947, Maurice Fargues—who had tended the ropes for Cousteau and Dumas in the Fontaine-de-Vaucluse—attempted a new depth record. He descended rapidly, tugging on a safety line attached to his weight belt as he went down to let the men on the surface know he was alive, a method of communication that was standard for hard-hat divers. Fargues reached 385 feet in three minutes, scratched his initials on a slate he tied to the anchor rope to verify the record, and signaled that he was okay. Long seconds passed with no tugs on the line. Tailliez barked the order to haul him up. At 150 feet, another diver on the way down the anchor rope saw Fargues hanging limply from the safety line, his mouthpiece dangling free. On the surface, they worked for hours to revive him but to no avail. Fargues had been breathing ordinary compressed air. Almost certainly, Cousteau and the others concluded, he was the first scuba diver to be killed by rapture of the deep. His death, and the new depth record confirmed by his scribbled initials on the slate at 385 feet, made headlines across Europe.

The death of Maurice Fargues hit everyone in the group hard. Cousteau was especially distraught. His Aqua-Lung had killed its first diver. "Maurice had shared our unfolding wonderment of the ocean since the earliest days of the Undersea Research Group," Cousteau wrote in his memoir *The Silent World*. "We retain the memory of his prodigal comradeship. We will not be consoled that we were unable to save him."

Since the time Cousteau had suffered convulsions while testing the first oxygen rebreather seven years earlier, he had known that breathing compressed gas underwater would be dangerous in some unexpected ways. Every incident was above all else a problem to be solved. Fargues's handwriting on the slate at 300 feet was completely legible; at 385 it was a scribble. Apparently, 300 feet was the maximum depth a scuba diver could reach while breathing ordinary compressed air.

Cousteau and the other divers immediately began experimenting with mixing air and helium to replace the nitrogen and reduce the possibility of narcosis.

As a tonic to banish the grief and the implications of a diver dying while using an Aqua-Lung, the Undersea Research Group embarked on its first archaeology expedition, to a sunken Roman ship off the coast of Tunisia. Tailliez, who was still in charge of the group, gave Cousteau command of the former German dive tender *Albatross*. The 78-foot vessel had been seized by the Russians, then given to the British, and finally handed over to the French navy as the spoils of war were shuffled around Europe among the victors. The group renamed it *L'Ingénieur Elie Monnier* to honor a navy engineer and hard-hat diver who had died when a bomb detonated under him while he was inspecting the wreck of the battleship *Bretagne*. With it, they would be able to travel across the Mediterranean, live relatively comfortably, make dozens of dives to the wreck, and use the ship's cranes to haul up what they found.

As Cousteau directed the preparations for departure, he was very aware that he was developing a model for the self-contained diving expedition that, until then, had not existed. He planned for a fourteen-day voyage, constructing detailed lists of supplies and equipment, including food, wine, water, cigarettes, sidearms, rifles, and flare guns; Aqua-Lung regulators with spares for every part; air tanks; compressors; a one-man recompression chamber; still and movie cameras; pressurized camera housings; film, including some experimental color movie film; medical kits; the towed sled they had developed for minesweeping; cables, ropes, and baskets for hauling up artifacts; and hoses to carry compressed air to the bottom, with which Cousteau thought he might be able to scour sediment from the wreck. Cousteau's biggest challenge was stowing the mountain of equipment while leaving enough room for accommodations for his crew of thirty. *L'Elie Monnier* had a large, tugboatlike deckhouse but not much hold space, so he and his crew spent weeks building bins, racks, and compartments on deck to hold their diving gear.

The wreck of the Roman ship they set out to explore had been first discovered, coincidentally, in 1907 by Simone Cousteau's grandfather, Admiral Jean Baehme, after a Greek sponge diver told him he had seen giant cannons half-buried in the sand about 130 feet down.

Baehme had sent navy helmet divers down to investigate. The cannons turned out to be Athenian marble pillars typical of the first century B.C., one of which the divers hoisted to the surface with a crane on their tender. Baehme had called in an archaeologist who told him that the wreck was probably the remains of a ship that had been carrying an entire Greek temple or villa, most likely plunder from a Roman raid. The French navy turned the wreck over to a museum in Tunis, which salvaged enough to fill five of its rooms with the most sensational underwater artifacts ever found. In another coincidence, James Hazen Hyde, who was then employing Daniel Cousteau as his traveling companion, had contributed $20,000 to the five-year operation. It had been enough to salvage less than half the precious cargo. The rest was still there.

Cousteau's first stop in Tunisia was the archaeological museum, where he found sketches that pinpointed the wreck by triangulating landmarks on shore: a castle, some odd vegetation in the sand dunes, and a windmill. The next day, as *L'Elie Monnier* cruised a few hundred yards off the coast in the vicinity of the wreck, Cousteau peered through binoculars and discovered that thirty-five years of wind, erosion, and shifting sands had spared only the castle. One landmark was worthless, but Cousteau still had two solid facts. He had seen the depth-sounding records from 1907 and knew that the wreck was at 127 feet, and he knew from rough positioning notes on the chart that it was somewhere nearby.

At the center of the estimated position of the wreck, divers laid out a steel wire grid covering 100,000 square feet, then systematically searched every inch of it. According to the decompression tables, a diver could stay at 130 feet for fifteen minutes and surface without stopping. To mark time during the dives, a rifleman on the surface fired a shot into the sea every five minutes, and two shots at fifteen minutes. After each shot, the divers instinctively looked up and watched the rifle bullets sink harmlessly to the bottom through crystal clear water, with the hulking shadow of *L'Elie Monnier* above them. They dove all day in shifts of two divers, each man making no more than three dives, but found nothing but sand, gravel, and occasional debris that slowed their search and proved to be meaningless. The next day, they towed divers on the sled in ever increasing concentric circles around the estimated position at a depth of 100 feet, from which they could see the bottom but remain in the water for almost a half hour without decompressing. Still nothing.

On the fifth day, as Cousteau and his divers grappled with the demoralizing possibility that their expedition would be a failure, Tailliez volunteered to be towed behind their auxiliary launch without the sled to take a wild stab at getting lucky. Tailliez made his offer to break the tension by taking all the responsibility for success or failure on his own shoulders as the group's commanding officer. Every other diver also volunteered. The following morning they began a series of twenty-minute shifts that went on fruitlessly until late in the day. Tailliez went down for the last of his three dives at about four thirty in the afternoon, searching in an arc about 700 yards off the headland that was a half mile from where they had expected to find the wreck. Ten minutes into the dive, Tailliez glided over what was unmistakably a marble column, then another, and another. They were splayed like jackstraws around the clearly discernable outlines of a ship's hull.

After finding the wreck, they had just a week to explore it before their navy orders required them to be back in Toulon for another assignment. Cousteau's scouring air hoses worked beautifully. Though they sent a plume of muck, bubbles, and rocks toward the surface, forcing the divers using them to lie flat on the bottom beneath the debris, the hoses accomplished in minutes what would otherwise have taken hours. The wreck, which only vaguely resembled a ship, was 130 feet long by 40 feet wide—twice the size of *L'Elie Monnier*—lying on a plain of sand and mud, swarming with fish whose numbers increased as the hoses tore organic debris from the bottom. Before the first dive, Cousteau and the others had agreed they would not spear any of the fish, chiefly to keep down the possibility that sharks might be attracted to the blood. After a single day of diving on the wreck, the fish had become so comfortable with the giant but harmless invaders that they swam a few feet away from the divers to snatch the first morsels of food stirred up by the hoses.

For six days, rotating in two-man teams, the group worked from sunup to sundown, controlled by the "rifle clock" from above. Only Dumas, who could not resist three extra minutes on the bottom after he spotted something shiny on the wreck at the end of his third dive of the day, got bent. He seemed fine until he sat down to dinner complaining of a slight pain in his shoulder. Cousteau and two other men jumped up and pulled the grumbling Dumas onto the deck and into the recompression chamber. They set the depth to 4 atmospheres—132 feet—and locked the door. The chamber was equipped with a telephone connected to a loudspeaker in the galley, through which Dumas kept up a steady banter, accusing his fellow divers of starving him as a prank. All of them knew, though, that they could not take even the slightest chance with the bends.

Conditions were perfect for testing some new color film Cousteau had gotten from the French Film Institute. In the dim blue light at 130 feet, he shot the faint yellow sunlight winking off the divers' chrome regulators, with their bubbles streaming behind them like capes of shimmering jewels. The divers themselves, swimming only in brief bathing suits, were putty-colored and ghostly. The bright sand bottom reflected enough light to illuminate the fluted marble columns, capitals, and bases as the divers attached cables and maneuvered them into position for their return to the surface after two millennia on the bottom of the Mediterranean.

Cousteau and his men were delighted with the efficiency of the world's first industrial salvage operation using the Aqua-Lung, and awestruck by ancient artifacts from the wreck. Iron nails corroded to the thickness of needles, bronze nails worn down to wires, and still-varnished ribs of Lebanon cedar revealed the ship to them. They recovered a millstone used by the ship's crew to grind grain stored for the voyage in the amphorae they found intact in the center of the wreck where a hold might have been. They found and raised two 1,500-pound iron anchors. With ropes and cables, they brought four 3-ton columns to the surface, scrubbed them off, and marveled at the white marble that emerged from the slime. The stone clearly showed the chisel marks made by a carver more than two thousand years before, when the Mediterranean was the center of the known world. Except for a few artifacts kept by the divers as souvenirs—including an Ionic capital that Cousteau took home with him to Sanary-sur-Mer—everything recovered by the first international archaeological expedition of the Undersea Research Group remained at the museum in Tunisia.

"The finest treasures in the world are waiting in the Mediterranean, now within range of the lung," Cousteau boasted in an interview after the Tunisian adventure. "She is the mother of civilization, the sea girt with the oldest cultures, a museum in sun and spray."

In the most frequently published newspaper photograph from the

expedition to Tunisia, Guy Morandière is shown guiding a spectacu-
larly detailed Ionic capital slung in a rope cradle from a crane that is out
of the frame. Sunken treasure had always been sensational fare for writers and reporters. The possibility that archaeological stories from the
ancient past could now be told by divers who could breathe and swim
free underwater sent a minor shock wave through postwar Europe,
hungry for reassurances that humanity was back on the right track.

Among those who noticed the tremor was an American reporter
for *Yank* magazine, James Dugan, who had met Cousteau shortly after
VE day, when Cousteau was in London trying to interest the British
and American navies in his Aqua-Lung. Dugan had never wanted to
be anything but a writer. He was born to middle-class parents in
Altoona, Pennsylvania, in 1912, and took advantage of the peaceful
years between the world wars to graduate from high school and
become a member of the class of 1935 at Penn State. It was an era in
America during which—thanks to Hemingway, Fitzgerald, and Walter Winchell—becoming a writer, reporter, or radio announcer
could make a young man popular and attractive. Dugan parlayed a little talent into the editor's chair of the college literary magazine, then
went to New York to write articles for *Holiday, The Saturday Evening
Post,* and the flood of magazines that made newsstands the center of
the information world.

After serving for two years as an enlisted man teaching language
classes to soldiers while the United States sat out the beginning of
World War II, Dugan shipped out to England as a combat correspondent with the Army Air Corps. He flew missions over the Ploesti oil
fields in Romania, wrote good stories, and later found a job as a reporter
for *Yank* magazine. In early 1945, with the Allies closing in on victory,
Dugan was in London, where he saw a short film of men swimming
around a shipwreck wearing tanks of air on their backs. He was fascinated and so sure his friends and editors would not believe what he had
seen that he went back to the newsreel theater with a still camera and
clicked off a couple of frames. Dugan noted that the credits of the film,
Épaves, listed Jacques-Yves Cousteau as director and cameraman and
knew that he had to find this brave or foolhardy Frenchman who had
figured out a way to swim underwater like a seal.

The following spring, Dugan heard that Cousteau was in London,
prowling the hallways at the admiralty with photographs and sketches

of his underwater breathing apparatus, and tracked him down for an interview. They met for breakfast and talked through lunch and dinner. Dugan was charmed by the Frenchman's enthusiasm and sense of purpose. Cousteau was not a man who would waste much time with anyone who did not fit into his plans. After dinner, Dugan spent what remained of the night typing out his notes, which became a long magazine article, "The First of the Menfish," betraying the writer's enthusiasm for his scoop and his instant attachment to Cousteau:

> In the clear, warm waters of the French Riviera a new species of large mammalian fish have been observed in the last few years, one-eyed monsters shaped and colored like nude human beings with green rubber tail fins, gills of metal, and tubular scales on their backs. They are called Cousteau Divers. They swim around sportively at hundred-foot depths, examining sunken ships, taking photographs, and harpooning big fish. They are the first of the menfish, a new order of marine life invented by Lieutenant de Vaisseau Jacques-Yves Cousteau of the French Navy.

In ten thousand words, Dugan methodically traced the evolution of the new aquatic species: the invention of the Cousteau-Gagnan valve, which delivered a measured breath of air regardless of the diver's depth or attitude in the water; the pioneering Aqua-Lung divers' flirtations with death as they tested the limits of their invention and experienced the dangers of breathing gas under pressure; Cousteau's obsession with making movies underwater.

"The entire motive of the fifteen years of work Cousteau has put into his lung was to make films under water. He has the movie bug bad," Dugan wrote. "Lieutenant de Vaisseau (gunnery officer) Cousteau is 37, a lean, bronzed man with a bold Mediterranean profile and eyes of ocean blue. He bought his first movie camera when he was 13 . . ."

After Cousteau finished a couple of years of scientific work, Dugan wrote, he planned to make a romantic feature film under the ocean. As fantastic as it might sound, Cousteau said he would have his diving technique worked out by this time so that the stars would be able to swim around 50 feet down without face masks or air bottles.

"I will show you a beautiful mermaid swimming around with the ghost of a drowned pirate for fifteen minutes without masks," he promised. "No studio tanks, no fake optical printing. I will film it in the sea."

Dugan sent his story to editors in London, New York, and Paris. All of them rejected it.

Within three years, the exploits of Cousteau and his band of Aqua-Lung explorers had begun to creep into newspapers and newsreels. Dugan sold his story in early 1948 to *Scientific Illustrated,* a small magazine edited by a friend from his days as a war correspondent. A selection of black-and-white photographs, printed from frames of 35 mm movie film, appeared with the story. One shows Dumas playing with an octopus and chasing a giant manta ray along a reef. Another shows four Undersea Research Group divers descending into an underwater cavern, looking like shadows against the light from above them. Tailliez, armed with his speargun, takes aim at a target just to the right of the camera's position in another shot, and a final frame shows the ghostly remains of the ship's wheel on *Dalton,* surrounded by a halo of fish, clearly oblivious to the intruder taking their picture.

Almost as an aside, Dugan had written that the revolutionary Cousteau Aqua-Lung would soon be available in the United States and Canada through a subsidiary of Air Liquide.

Émile Gagnan had immigrated to Canada in 1947, fearing that France after the war was bound for a future as a Communist country. With the help of Canadian Liquid Air, a subsidiary of Air Liquide, for whom he still worked, Gagnan set up his laboratory and workshop in Montreal to produce Aqua-Lungs for the North American market. Before Gagnan left France, Air Liquide formed a new division it called Spirotechnique, a stand-alone corporation in which it owned 99 percent of the shares. Cousteau owned the other 1 percent. In creating Spirotechnique, Air Liquide was particularly careful about constructing a legal barrier between the assets of the new corporation and its own massive wealth. While everyone was enthusiastic about entering the robust North American market with the Aqua-Lung, no one was certain that the revolution they had pioneered with highly trained, very fit military divers could be safely extended into the civilian world.

After Dugan's article on the menfish was published, requests from readers wanting to know where they could buy an Aqua-Lung poured in to the editors of *Scientific Illustrated.* The only source for scuba gear in the United States, they wrote back to their readers, was Spaco, Inc., a foreign manufacturing representative with its headquarters in Vermont and offices in New York. Spaco had a deal with Air Liquide for the rights to import and distribute the Aqua-Lung on the East Coast of North America. They sold the few they were getting from Gagnan in Canada to dealers, including the internationally known sporting outfitter Abercrombie and Fitch. On the West Coast, Air Liquide was negotiating rights with an expatriate Frenchman in Los Angeles, René Bussoz.

For the better part of a year, the news that the Aqua-Lung was for sale in North America found its way into military strategy sessions and waterfront gossip on both coasts. An American admiral who had made an Aqua-Lung dive at Cannes shortly after the war staged an impromptu competition between a scuba diver and a hard-hat diver clearing wreckage from the edge of the shipping channel in Norfolk, Virginia. The diver wearing an Aqua-Lung easily outperformed the hard-hat diver, who was bedeviled by strong currents. The admiral was happy to open the door for sales to the U.S. Navy experimental diving team.

9

THE ABYSS

WITH JAMES DUGAN'S ARTICLE on menfish in print and the demand for Aqua-Lungs outstripping supply, Cousteau was beginning to realize that he was no longer an obscure French gunnery officer with a passion for diving and filmmaking. He did not intentionally seek the spotlight, but when illuminated, he instinctively performed with a charming blend of forthrightness, humor, and an unself-conscious, boyish curiosity that was irresistible to reporters, news cameras, and admirals.

"My father loved his life, especially after surviving the war," his son Jean-Michel said. "His curiosity and ambition were starting to pay off and everything seemed like playing to him."

There appeared to be no limits to the contributions Cousteau and the Undersea Research Group were going to make to the world's knowledge of the sea. Tailliez remained in command, but he was still the shy, stammering man Cousteau had met before the war. He was infinitely competent, but much happier orchestrating the work of the group from the wings, while Cousteau took the bows. In the four years since *Épaves* had convinced the navy to embrace and support the group, Cousteau and Tailliez had built their fiefdom on the Toulon waterfront. They had a three-story building on a wharf where *L'Elie Monnier* and an assortment of smaller diving tenders were moored. On the ground floor of their headquarters, they had a machine shop, photo lab, generator room, compressor room, and one of the few decompression chambers in Europe. On the second floor were the drafting room and crew's quarters. Cousteau and Tailliez had their offices on the top floor, with physics, physiology, and chemistry labs. Their conference room doubled as a museum for some of the booty the group had recovered from the sea: ships' bells and wheels, amphorae from the Roman wrecks, and pieces of marble columns from the Tunisian expedition. Most impressive was a water-filled tube running from the first floor through to the roof, in which the group could simulate water pressures down to 800 feet.

"Other navies had larger diving centers," Cousteau remembered.

> But ours had the virtue of an intimate connection with the sea beneath our windows . . . If Dumas had a new undersea device, the draftsmen and machinists could have a model ready for tests the next day. Where there were diving accidents, naval or civilian, the emergency patients were brought to our doctors. Bent and agonized men were carried into the physiological lab for decompression and we would all have the pleasure of gathering to see the patient emerge leaping and rejoicing.

Even with orders for equipment flowing in from navies, salvage divers, and scientists, Cousteau was surprised and honored when Swiss inventor and explorer Auguste Piccard came to Toulon to ask for help building a diving machine he said could reach the eternal darkness of the abyss. "Professor Piccard, who had been eleven miles in the sky, now proposed to go thirteen thousand feet into the ocean," Cousteau remembered. He called Piccard a "scientific extremist," in a tone of humorous admiration.

Piccard was the most celebrated pioneer in a scientific community whose chief preoccupation at the time was determining the limits at which life could sustain itself. A physicist by training, he had been an unknown though well-regarded professor in Brussels until 1930, when he turned his attention to the composition of gases and effects of cosmic rays in the earth's upper atmosphere. The best way to prove his several hypotheses was obvious: go there. With funding from the Belgian national science foundation, Piccard built a pressurized aluminum gondola big enough for two men and slung it under a gigantic helium balloon. On a still, spring morning in 1939, he took off from a farmer's field near Augsburg, Germany, with his assistant Paul Kipfer. Seventeen hours later, their gondola and the partially deflated balloon, having reached an altitude of 51,775 feet, crashed down on a glacier in Austria. Both men survived. Piccard broke his own record the following year, and after twenty-seven more balloon flights

Auguste Piccard's deep-diving bathyscaphe (ACTUALIT)

reached 72,177 feet, a record that would stand for thirty-five years. Piccard's ascents into the stratosphere made international headlines, and he quickly became a master of the art of turning his sensational exploits into sponsorships for his next earth-shaking expedition. In 1937, he announced that the same kind of pressure gondola he had used to set his altitude records would be perfect for exploring the depths of the ocean.

"It is a world as unknown to us as the surface of the moon," Piccard told reporters.

Piccard set aside his ambitions for exploring the ocean when Germany invaded Poland. He went back to work on the diving project after the war, and by the time he came to Toulon had completed the heavy steel capsule he called a bathyscaphe, which means "deep boat" in Greek. The first part of the name is an homage to the earliest pioneers of deep-ocean exploration, the celebrated naturalist William Beebe and his partner Otis Barton. From 1930 to 1934, inspired coincidentally by Piccard's revolutionary ascents into the upper atmosphere, Beebe and Barton had built a 4.5-foot steel chamber with quartz glass portholes that they called a bathysphere, meaning "deep sphere." In it, they had been lowered on the end of a cable into the Atlantic off Bermuda, reaching an ultimate depth of a half mile and becoming the first human beings to peer into the abyss. Barton had even succeeded in producing primitive moving pictures with a hand-held camera pointed through one of the portholes. Piccard rejected the bathysphere's cable in his own design, deciding to control his ascent and descent with lead ballast and enormous chambers of lighter-than-water gasoline—much as he would a helium balloon. Though his bathyscaphe would be capable of only slight course changes and speeds under one knot while submerged, the absence of the cable greatly simplified dives beyond the half mile Beebe and Barton had reached. The weight of a cable alone would be prohibitive, and the most terrifying risks for the occupants of a suspended gondola were broken or tangled cables that would doom them.

When Piccard came to Toulon with his blueprints for the bathyscaphe, he told Cousteau and Tailliez that it was worthless to risk their lives diving into the abyss and return with no samples. He wanted the Undersea Research Group to design a grappling claw to pick things up from the bottom and a battery of harpoon cannons to shoot fish and other creatures. They might even get lucky, Piccard said, and bring back the first specimen of the legendary giant squid. Piccard also wanted Cousteau and the other Undersea Research Group divers to come with him to West Africa to photograph the bathyscaphe in the water and help with the difficult business of launch and retrieval from the water. Finally, Piccard said that one man from the group could make a dive to test their equipment and photograph the abyss through the thick portholes of the bathyscaphe. The sheer adventure of it would have been enough for Cousteau, but the fact that Auguste Piccard had come to him for help made the challenge impossible to turn down. The Aqua-Lung had opened a new universe to exploration, but it had already clearly defined its limits. Cousteau longed to go farther into the depths of the ocean. He assumed that he would be the man to make a dive with Piccard in the bathyscaphe.

Two weeks before the Undersea Research Group was set to sail for West Africa, Cousteau's leg was in plaster. Despite finding his natural athleticism in the water, he had continued to avoid physical games unless his sons were involved. Jean-Michel was a natural at anything to do with hand-to-eye coordination and had especially taken to tennis. On the city court in Sanary, Cousteau had lunged for the ball, fallen awkwardly, and broken a bone in his foot. Simone was sympathetic about her husband's injury, but secretly delighted that he might not be able to go on the bathyscaphe expedition.

"No one ordered you to go," Simone said when she broached the subject of JYC staying home. "Don't risk yourself in that craziness. Please don't go down in that horrible machine."

"It was one of the rare times Simone had objected to any of my plans," Cousteau recalled. "She was a navy wife with a self-disciplined attitude toward my activities."

Simone wasn't the only member of Cousteau's circle of family and friends to urge that he steer clear of Piccard and what they cynically called the "submarine dirigible." Both Daniel and Elizabeth begged him to stay home to let his foot heal. Several of the scientists who worked with the Undersea Research Group cautioned him against allying himself with a man considered to be a daredevil showman. They said he had been a distinguished physicist early in his life, but now seemed to be more interested in making headlines and raising money than in real scientific inquiry.

There was nothing reckless in Cousteau, and he tried to ease their concerns. "The bathyscaphe is perfectly safe," he told them. "There is nothing to worry about." But even if he were concealing uncertainty about his chances for success, he was determined to go with Piccard. Anything as bold as descending into the abyss where only a handful of people had been before him was bound to be dangerous, but the rewards far outstripped the risk. Piccard's bathyscaphe was a primitive vehicle, but Cousteau was sure it was only the beginning. He wanted to be part of the adventure.

"It was impossible to resist," Cousteau admitted. "Tailliez, Dumas, and I were together again, about to sail to West Africa on our greatest adventure, and nothing could stop us. I had been selected to enter a wonderful submarine dirigible and dive five times deeper into the sea than man had ever gone."

Scuba divers had worked at depths down to 400 feet. Military submarines cruised only slightly deeper. Beebe and Barton had been to a half mile, where they remained for only minutes. Piccard's bathyscaphe, if it worked, meant human beings would be able to investigate the sea and its creatures down to 13,000 feet, a depth that allowed access to almost 90 percent of the ocean floor. Only the deepest trenches in the midocean rifts were deeper.

From the blueprints of Piccard's chief designer, Max Cosyns, Dumas and the engineers at the Undersea Research Group had spent more than a year building the grappling claw and harpoon cannons, each of them utterly unique. The claw was a simple mechanism mounted on an arm outside the bathyscaphe. The engineering problem was finding a way to control it without breaching the pressure hull, which had to remain intact under pressures as great as 6,000 pounds per square inch. Dumas and his team solved it with a two-piece fitting that went through the hull, one piece of stainless steel and the other of aluminum. The metals of differing degrees of hardness compressed when tightened to seal the hole around the control wire leading to the grappling claw.

The harpoon cannon they built was controlled by wires through similar fittings, but it was far more bizarre and ingenious than the grappling claw. Each of seven tubes in the cannon carried a .25 caliber three-foot-long harpoon propelled by the immense pressure that would build up in a chamber at the base of the cannon as the bathyscaphe descended. At a depth of 3,000 feet, the harpoons could pierce three-inch-thick oak planks; on the surface, they were harmless. The harpoons were designed for taking specimens of abyssal creatures— the giant squid being the grand prize—so each was tipped with a strychnine reservoir that would burst on impact. If that wasn't enough to make the kill, the bathyscaphe hunters cold trigger an electrical discharge to further immobilize their prey. At the base of the cannon were seven reels with lines attached to the harpoons for reeling in the speared specimens.

Cousteau loved the claw and cannon, which reminded him of the wondrous technology of Captain Nemo's fictional submarine *Nautilus* in the films of Georges Méliès and the Williamson brothers. The action in *20,000 Leagues Under the Sea* had featured battles with sea monsters that dominated the imaginations of everyone who saw the films. If there really were giant, deadly creatures down there, Cousteau would be ready for them.

On October 1, 1948, Cousteau took *L'Elie Monnier* out of Toulon. He set a course past Gibraltar and south to the coast of Dakar to rendezvous with the Belgian freighter *Scaldis* carrying Piccard, the bathy-scaphe, a half-dozen scientists, and a troop of journalists. Tailliez, Dumas, and everyone in the Undersea Research Group not absolutely needed to stand duty at their headquarters were aboard *L'Elie Monnier,* along with the grappling claw and harpoon gun. Théodore Monod, director of the Institute for Black Africa, and oceanographer Claude-Francis Boeuf sailed with them as scientific observers for the French government. The navy sent two frigates and a reconnaissance plane to accompany the mission that would enhance the image of France as a power in ocean exploration around the world.

As soon as *L'Elie Monnier* was at anchor off the coast of West Africa, Cousteau, Tailliez, and Dumas went over to *Scaldis* to inspect the bathyscaphe, which they had not seen until then. The strange craft looked enormous resting in its special cradle in the midships cargo hold on the freighter. Cousteau was startled to see that it looked even more like a dirigible than it had in the blueprints. The top was a diamond-shaped metal balloon 30 feet long and 15 feet high, under which was suspended a steel sphere about 6 feet in diameter to carry two passengers. In each of two hatches on the side of the sphere was a glass porthole. Cousteau knew from the plans he had studied that the balloon held six steel tanks that could carry a total of 2,500 gallons of gasoline for buoyancy. Coiled under the sphere was a cable, at the end of which was a huge weight shaped like the blade of an ice skate. The bathyscaphe also carried more than a ton of iron-shot ballast that could be released. Descending and ascending were theoretically sim-ple: The lead and iron weights took the bathyscaphe down when some of the gasoline was pumped out of its tanks. When some or all of the weight was released, the craft either hovered in midwater or rose to the surface from the buoyancy of the remaining gasoline.

"My trust in the bathyscaphe was reinforced when I saw it," Cousteau said. "I knew the principles by heart from the blueprints, and now I touched the real thing."

When the moment came to select a member of the Undersea Research Group for the first test dive, Cousteau, as *L'Elie Monnier*'s captain, typically insisted that he and his men draw straws for the honor. Even if he lost the drawing, he was confident that he would be able to make one of the all-up test dives into the abyss to test the grappling arm and harpoon cannon. Théodore Monod, the director of the Institute for Black Africa, won. On the afternoon of November 26, Piccard and Monod were sealed in the sphere. The freighter's cargo crane lowered the bathyscaphe into the water for the dangerous business of pumping 2,500 gallons of gas into the tanks. The ship's crane was not powerful enough to lift the gassed-up and ballasted sphere, and in case something went wrong, an explosion in the water would do less damage than an explosion on the ship. With Cousteau hobbled in his plaster cast standing on deck, Tailliez and Dumas led a team of safety divers to check on Piccard and Monod in the suspended sphere. Tailliez surfaced, flipped up his mask, and yelled to Cousteau, "Everything is okay. They're playing chess."

After nightfall, the bathyscaphe finally slipped beneath the sea, surrounded by a glowing corona from its lights that dimmed to a bright haze as it descended to only 200 feet for the first test dive. Sixteen minutes later, Cousteau watched as the sea brightened and the top of the balloon broke the surface. On the ship, cheers erupted but the celebration didn't last long. For five long hours, with Piccard and Monod sealed inside the sphere, *L'Elie Monnier*'s divers pumped gasoline into the sea. Finally, the crane could lift the bathyscaphe. On deck, with movie cameras rolling under floodlights, the crew opened the hatches to free the exhausted men inside. Cousteau never forgot what he saw.

"A high leather boot came out, followed by a bare shank, another boot and leg, bathing trunks, a naked belly, and the bespectacled wild-haired pinnacle of Professor Auguste Piccard," Cousteau wrote later. "His hand was extended, clutching a patented health drink with the label squarely presented to the cameras. Professor Piccard ceremoniously drank the product of one of his sponsors. The bathyscaphe was back from the deep."

Cousteau's delight at watching the world's greatest science showman perform at the end of a grueling dive was replaced two days later by abject disappointment. During the next test, an unmanned descent to 4,600 feet over an undersea canyon, the thin metal balloon holding the gasoline chambers was bent and crumpled beyond repair. After struggling through the night to release ballast and pump off gas, divers finally lightened the bathyscaphe enough to bring it out of the water, but there would be no more diving on that expedition.

Cousteau was crestfallen that he would not make a dive to test his grappling claw and harpoon cannon, but he also knew that Piccard was not going to stop working on the bathyscaphe. It was only a matter of time before he and his engineers made one that worked. He turned *L'Elie Monnier* homeward for Toulon also knowing that joining Piccard on his first expedition to explore the abyss signaled his acceptance into the top rank of ocean exploration.

After the publicity from the bathyscaphe expedition and the rest of the attention Cousteau was getting from the press, he hired his father as a full-time business agent. Eugene Higgins had died leaving Daniel without a job, so he was happy to divide his time between Sanary-sur-Mer, Paris, and Torquay. After his first trip to Sanary, he was known simply as Daddy because of his command of English, his relationship with JYC, and the genial, fatherly presence he brought to every situation. He had an agent's soul, with great intuition for who was worth his son's time and who was not. He gracefully deflected those he decided were not part of the way forward, aware that one never knew if they would be of use to his son's career later on.

One of the first introductions Daddy brokered for JYC was with a young woman named Perry Miller, a United Nations cultural attaché scouting in Europe for new films to feed the energetic postwar documentary boom in the United States. Miller was pretty, adventurous in the style of a Hemingway heroine, and independent at a time when magazine ads were celebrating housewives in bouncy petticoats.

When Miller met Cousteau during a visit to Paris, she instantly recognized an animal spirit in him, a magnetism that she and almost every other woman he had ever known would find impossible to resist. She knew that not every attractive woman JYC charmed would become his lover, but there was no question that in the first two minutes after meeting him most would decide, at least theoretically, that they would. Smitten by Cousteau and convinced that he radiated the unmistakable scent of stardom, Miller went back to New York determined to help Americans fall in love with this brave, charismatic Frenchman. Daddy gave her prints of *Par dix-huit mètres de fond, Épaves,* and two other short films, one from the Tunisian expedition, the other a navy training film about escaping from a sunken submarine using a breathing lung. At a gala evening in New York in early 1950, Miller premiered Cousteau's films, along with several others from her European trip. The next day, a *Life* magazine editor who had been at the screening called to ask if he could take another look at the ones of the French fellow who breathes underwater.

In November 1950, *Life* ran a seven-page spread of photographs. Most had accompanied Dugan's *Science Illustrated* story, but in *Life* they were printed many times larger and seen by many more people. Under the headline "Underwater Wonders," Dumas wrestled an octopus, the Undersea Research Group team swam downward from the light above and into the depths, and shipwrecks once thought lost forever came alive again. In an inset photograph on the first page, Cousteau and his underwater movie camera evoked a scene straight out of Jules Verne. The article ended with five increasingly large close-ups of a menacing shark. The final caption, under an image of a shark's gaping mouth filling the frame, was "Cousteau bumped the shark's head with his camera and got this frightening nose-to-nose close-up. The shark retreated, and divers rose as quickly as possible." Cousteau was quoted as saying he had seen forty-three sharks in a single month of diving. None of them showed the slightest inclination to attack him, probably, he said, because he carried cupric acetate as a shark repellant.

The rest of the *Life* issue, which had a paid circulation of more than ten million and a readership estimated at five times that number, featured the UCLA homecoming queen Allyn Smith on the cover, news reports on a plane crash in the Alps that orphaned nineteen children in a single Canadian family, a U.S. jet fighter pilot shooting down the first Russian Mig over the Yalu River that divides China and Korea, the assassination of Venezuelan dictator Delgado Chalbaud, and the murder of five family members by an estranged husband in New Jersey. The photo essay on Cousteau and his divers led the features section, followed by articles on the annual celebration of the Marine Corps birthday; a new discovery about the cause of high blood pressure; hair dyes that could be used safely at home; and the mysterious British billionaire, Calouste Sarkis Gulbenkian, his art collection, and his control of the oil exploration in Arabia.

Niblets Sweet Corn bought the ad on the back cover of the issue, and I. W. Harper Kentucky bourbon the inside back cover. On the inside front cover, the Forstmann Woolen Company ran a full-page illustration of an elegantly wool-clad model against backgrounds of yellow, purple, and red fabric. Other products touted in the issue included General Motors airplane engines, Lucky Strike cigarettes, Schlitz and Budweiser beer, Playtex Fab-Lined girdles, Revere movie cameras and projectors, and television sets by Zenith, Spartan Town and Country, General Electric, Truetone, and the Capehart-Farnsworth Corporation. On television, Americans were watching *What's My Line?*, *Your Show of Shows*, *Hawkins Falls*, *Truth or Consequences*, *The Jack Benny Show*, and *The George Burns and Gracie Allen Show*.

The day after the issue of *Life* hit the newsstands, Perry Miller got a call from someone who said he was from Universal Pictures in Hollywood. Who did he have to talk to about rights to the underwater movies he had read about in *Life*? Miller referred him to Daniel Cousteau. A week later, Jacques Cousteau accepted Universal's offer of $11,000 for exclusive U.S. rights to his first four documentaries.

10

CALYPSO

AFTER FOUR YEARS OF expeditions aboard *L'Elie Monnier*, Cousteau wanted a new ship. He remained grateful for the converted German patrol boat that had been his first command but it was just too small and poorly equipped to take him into the future. *L'Elie Monnier* had nowhere near enough stowage, a single engine that made it clumsy in close quarters, and limited deck space, which made launching and recovering teams of divers a nightmare. He wanted a bigger boat, with twin engines, a shallow draft for working around reefs, and accommodations, fuel, water, and cargo capacity for months instead of weeks.

At a meeting with the admiral in charge of the Undersea Research Group, Cousteau stood at attention and delivered the precisely worded request that he, Simone, Tailliez, and Dumas had written out the night before after dinner in Sanary-sur-Mer.

"Our team is ahead of everybody in putting a man in the sea," Cousteau said. "The national interest is to keep us in the lead with a new type of undersea research vessel built to the special needs of Aqua-Lung divers."

With a trace of sympathy for what he apparently took to be a naive and grandiose plan, the admiral turned Cousteau down flat. "As a lieutenant commander, you have no chance of getting a vessel. My advice is to return to routine duty. Work for advancement," the admiral told him. "Become an admiral. Then you might get your ship."

The following day, Cousteau went to another admiral who was even less sympathetic than the first. Not only was giving him his own ship out of the question, but since Cousteau had served seventeen of his eighteen years in the navy assigned to sea duty, it was time he did some staff time ashore.

Cousteau snapped to attention in front of the second admiral's desk. "With your permission, sir," Cousteau said. "From now on, I have one goal—to give my country an undersea exploring vessel. I request three months' furlough to look after personal affairs."

The admiral shook his head. Cousteau, he said, you are almost forty years old, no longer a young man. You have a solid career in the navy. This will ruin you, but I will grant you the furlough as I would to any other officer seeking to begin a business in civilian life.

Cousteau had no idea how to build a business around the exploration of the world's oceans, but banks were pouring money into the reconstruction of cities and the resurrection of industry, so he thought his timing might be right. He convinced naval architect Andrew Mauric to design the vessel he wanted gratis, took the plans to the French National Bank with the support of the French Film Board, and came away with enough money to make a movie but not to build a ship. Cousteau left the disappointing meeting with the bankers intent on borrowing enough from well-heeled friends to finance the ship, but not feeling especially optimistic about his prospects.

Cousteau's frustration increased after a succession of contacts from his father, Daniel, failed to pan out. Putting money into a new business built around a band of sailors who had learned how to breathe underwater was not attractive compared with the construction, manu-

Calypso (PRIVATE COLLECTION)

facturing, and production of consumer goods. Through Simone's father, Cousteau approached the board of Air Liquide, his partners in the promising but still nascent Aqua-Lung business. They offered only to supply tanks, compressors, and other equipment once Cousteau found his ship. Cousteau pointed out that movies, books, and news stories of the exploits of a team of underwater explorers using Aqua-Lungs was going to sell more Aqua-Lungs. Still no.

Dumas and Tailliez weren't having any luck either. A few friends said they might contribute to an expedition but none could stand the cost of an entire ship. At a particularly dark moment, when Cousteau was thinking he would be better off staying in the navy, Simone reminded him of an encounter with a British couple they had had at the bar of the Alpine ski resort at Auron after fleeing Paris in 1940. The woman had offered Simone half of her last cigarette—a grand gesture because tobacco was scarce. The couples segued from drinks into dinner, during which they talked about nothing but their shared passion for the ocean, their mutual fascination for the fantasies of Jules Verne, and the possibility that they were living in a time when those fantasies might become real. When they parted, the Cousteaus and their new friends agreed to get in touch after the war if their dreams of exploring the ocean survived.

Incredibly, Simone found the name, address, and telephone number of the couple from Auron in her address book. The man answered the phone as though he had been waiting for the call. When Cousteau told him he had developed the Aqua-Lung, explored the Mediterranean with it for four years, and was trying to raise money for a ship that would open the oceans of the world to him, the man invited them to come to London as soon as possible. He had a wealthy friend, Loel Guinness, who might be interested in making a contribution.

Loel Guinness, a lawyer descended from an Irish goldsmith, was the younger brother of the founder of the famous brewery in Dublin. He had inherited a fortune from his father, and lived the life of an international socialite until the war began. Then he had learned to fly, joined the Royal Air Force, and flew in the Battle of Britain. He was looking forward to a postwar life as a gentleman adventurer. In London, Cousteau told Guinness about the bad luck of the automobile accident that had ended his aviation career but then opened the way for his current fascination with the sea. He talked about the drop in the numbers of fish and the deterioration of the bottom of the Mediterranean off the south of France, and wondered aloud if the ocean, long thought to be impervious to destruction by man, might be more fragile than everyone believed. Guinness said that he, too, was obsessed with the possibilities that would arise from the exploration of the seven-eighths of the earth's surface covered by the sea. Apart from the coming to grips with the human impact on the sea, both men agreed that a dedicated research ship would no doubt open the ocean to exploration and enable the exploitation of the vast mineral and petroleum resources it contained.

For Cousteau, asking for money from a man he didn't know was tantamount to shameless begging. Simone, however, insisted that enlisting Loel Guinness in the grand adventure of underwater exploration was really offering the likeminded Irishman a chance to enjoy his wealth while contributing to scientific enlightenment. She was right. After he and Cousteau spent the better part of a day getting to know each other, Guinness said he believed the Aqua-Lung and a ship equipped to tend divers and produce movies could change the world.

Instead of building a new ship, however, Guinness had a better idea. Why not buy a much bigger, better ship from among the thousands being offered for sale as war surplus? In Malta alone, Guinness knew of a fleet of 115-foot British Fairmile-class torpedo boats in perfect condition and available for a few thousand dollars. Go to Malta, pick out a Fairmile, and I'll loan you the money to buy it, Guinness said. Cousteau was stunned by the generosity, but said he had no idea when he could repay him. Guinness said he wasn't worried about getting repaid. If Cousteau found the right ship, Guinness said he would lease it to him for one pound a year in perpetuity. He put two conditions on the lease. Cousteau could tell no one but his wife who paid for the boat, and he could never again come to Guinness for money. Like so many other people who fell under the spell of Cousteau's charm and ambition, Guinness was willing to help him, but as a businessman, he knew where to draw the line.

Two weeks later, Cousteau flew to Malta with naval architect Henri Rambaud to look at ships. He chartered a launch, cruised among dozens of anchored Fairmiles in the harbor at Valletta, studied their high-speed lines, and went aboard one of them to inspect the accommodations, deck space, engines, and stowage. Too small, Cousteau thought. Even with Guinness's blank check in his pocket, he didn't want to spend money on a boat that was less than ideal for tending scuba divers and underwater film crews.

On his way to shore, resigned to returning unsuccessfully to France, Cousteau spotted a much larger boat he recognized from his own navy service as a minesweeper. It was about 130 feet long, with a big, low afterdeck that would be perfect for diving, a solid wood hull, and, he could tell from the depth markings on its bow, a relatively shallow draft. Cousteau could see that unlike the idle Fairmiles, the minesweeper was in service as some kind of ferryboat.

When he landed, Cousteau walked back to the loading dock and went aboard. Its captain and owner, a transplanted Greek fisherman, was happy to show him around. The boat revealed signs of neglect, but most of it was cosmetic—blistered paint, rust streaks down its gray hull from the metal fittings and rigging straps. The decks and hull looked good, the caulking between planks still tight and sealed. The bilges weren't completely dry, but Cousteau knew that every wooden ship leaked a bit and this one wasn't taking water beyond normal limits. The pumps were obviously working. The engine room was hardly up to military standards of order and cleanliness, but Cousteau could see that the engines, propeller shafts, and stuffing boxes were being well maintained. Some of the interior bulkheads had been removed to make room for passenger benches and cargo stowage, and the decks cleared for carrying a few cars, but he saw the potential for refitting the former minesweeper as a comfortable research ship.

The ship was one of a series of minesweepers built at the Ballard Marine Railways in the Norwegian section of Seattle, Washington, in 1942 and 1943. Before the war, the little shipyard had been turning out halibut and salmon boats during boom years, and staying alive with repair work when fishing was bad. As in every other small yard in the Pacific Northwest, the shipwrights at Ballard Marine were experts in building with wood. With abundant supplies of cedar, cypress, spruce, and fir in the coastal forests, there was no better place on earth to build wooden ships.

Minesweepers were built of wood simply because metal hulls would set off the magnetic triggers of mines by accident before their crews were able to detonate them intentionally. The hull of *BYMS-26*, its U.S. Navy designation, was made of Port Orford cedar, a type of wood especially prized by boat builders and mariners because it was easy to work, strong, and highly resistant to rot. Actually a member of the cypress family, Port Orford cedar was unique to the forests along the Pacific in Northern California and Oregon, rising in dense stands to heights of 200 feet. It was straight-grained, which meant that long boards could be milled from the logs, greatly simplifying bending and fastening planks for boat building.

With a double hull of one-inch-thick planks, 6-inch frames on 18-inch centers, and 8-by-8-inch deck timbers, *BYMS-26* was strong enough to survive all but a direct explosion of a mine or artillery shell. The ship displaced 270 tons, was 136 feet long, 24 feet 6 inches wide, and drew 8 feet of water when fueled and loaded. It had accommodations for a crew of thirty-two, with armament of one 3-inch cannon and two 20-millimeter machine guns. Two 800-horsepower General Motors diesels could drive it at a top speed of 15 knots.

The Ballard Marine Railway and a half-dozen other yards around the United States built eighty of the BYMS minesweepers, which were then given to Britain for use in defending the harbors of the empire from attack. In the Royal Navy, *BYMS-26* became *J-826*, according to the British numbering system for minesweepers. It was assigned to the port of Valletta, Malta, where it saw action during the frantic late stages of the German and Italian collapse in 1944. In 1946, the war over, the Royal Navy sold *J-826* to a fisherman who converted it to carry passengers, cargo, and cars between the islands of Malta and Gozo. He took off the cannons, guns, and the minesweeping gear, scrubbed the navy numbers from its bow, and renamed his boat in honor of the mythical water nymphs of the Mediterranean Sea: *Calypso.*

Calypso's owner told Cousteau that every boat in the world is for sale at the right price. He named his. The next day, Rambaud inspected *Calypso* and confirmed Cousteau's own assessment. Its hull and decks were seaworthy and in need only of minor repairs and paint. Its engines and auxiliary generator had quite a few hours on them and were due for overhaul, but they wouldn't have to be replaced. Cousteau cabled Loel Guinness in London; Guinness said buy it.

Cousteau returned to Toulon, where he asked for and was granted a three-year furlough from the navy. He retained his rank, would return to active duty in time of war, but was otherwise free from all military obligations. Dumas, who had been released from his brief

service in the army after the war, signed on with Cousteau and his
new ship. Tailliez wavered, first deciding to leave the navy, then
reversing himself. He loved and respected Cousteau but there was no
question that even as his superior officer, Tailliez was increasingly
overshadowed by his friend. Better, Tailliez thought, to continue to
command the Undersea Research Group, cooperating with Cousteau
but moving away from him to make his own way. Auguste Piccard
had sold his first bathyscaphe to the French navy after the failures during test dives off West Africa, and Tailliez would be in charge of refitting it using the same steel ball but improved gasoline and ballast
tanks. To lead the Undersea Research Group through a challenge like
that was much more attractive to Tailliez than leaving the navy to
work in Cousteau's shadow. Tailliez harbored no ill feelings, but he
knew that expecting Cousteau to restrain himself from taking command of everyone around him was as futile as asking a fish to walk.

Cousteau, Dumas, their wives and children, and newly hired engineer Octave Leandri went to Malta to bring Calypso to the shipyard
in Antibes. One of the engines ran a little rough, but otherwise the
ship was sound. Turning a war-worn minesweeper into an oceanographic research ship was going to be expensive, not the easy, simple
escapade of three passionate divers on another lark, but an under-
taking that would require all of their cash and more. Before work
actually began, Cousteau asked for financial advice from a friend,
Claude-Francis Boeuf, who had experience raising money for large-
scale expeditions.

Boeuf told Cousteau that every explorer he had ever known was
always short of cash. Exploration itself produced knowledge but rarely
a profit. Cousteau explained that his plans included producing movies,
writing books and articles, and, with the publicity they created, selling
Aqua-Lungs. Boeuf said the movies, books, and articles—no matter
how successful—wouldn't pay for the operation and maintenance of a
130-foot ship and its crew. He suggested that Cousteau form a nonprofit corporation that would own the ship, manage expeditions, and
receive the royalties from films, books, and articles. Most important, a
nonprofit corporation could accept tax-deductible grants from indi-
viduals, science foundations, and private companies.

The first grant to Campagnes Océanographiques Françaises—

French Oceanographic Expeditions—was *Calypso* itself, on the one-pound-a-year lease from Loel Guinness. The second was cash from the Cousteaus, who mortgaged their house in Sanary and sold some of Simone's jewelry to begin refitting the ship. Since the first diving expedition to the *Dalton* wreck, Cousteau had known that he had a gift for enlisting collaborators and raising money, but he was startled at his own success in the first few months after he formed COF. Marseille industrialist Roger Gary, who had attached himself to Cousteau during the *Dalton* expedition, wrote a big check and persuaded his brother-in-law, the Marquis Armand de Turenne, to match it. Dumas pitched in some of his own money and talked friends into getting in on the ground floor of the world's first private oceanographic research corporation.

Calypso came out of the shipyard in the summer of 1951. It still had the lines of a minesweeper, but beneath its sparkling coat of white paint it was a completely different ship. Below the waterline, Cousteau had added a bulbous bow with eight viewing ports for filming underwater that coincidentally increased *Calypso*'s speed by half a knot because the new shape created less resistance to water passing under the bow. There was room for one observer lying prone on a mattress in the bow, or two, Cousteau joked, if they were very close friends. Through the floor of the galley, he cut a hole through which divers could leave the ship from inside by climbing down a ladder inside a tube that extended through the bottom. The water rose in the tube only to the height of the normal waterline, and a nonwatertight door had to be closed over the outside of the tube only to streamline the hull.

On deck, Cousteau had installed a crow's nest on a square aluminum scaffolding in front of the wheelhouse, which gave an observer an additional 20 feet of elevation. It had a full set of steering and power controls, and also held the radar and radio antennas at the highest point on the ship. Aft of the deckhouse was a towing winch for retrieving sampling cables, dredges, and baskets of artifacts on archaeological expeditions, and a light davit for launching and recovering dive tenders, compression chambers, and other equipment. On the top deck, behind the wheelhouse, was another davit for lowering and lifting an auxiliary motor launch.

Calypso's interior compartments were laid out on four decks. The lowest was half the height of the other three, containing only tanks for

fuel and water. Above it, the aft section of the next deck was taken up
by twin eight-cylinder diesels, generators, a machine shop for servic-
ing the engines, electrical busses and panels, and a large cargo hold.
The front section of that deck contained a cold storage locker for supplies and scientific samples, berths for six people, dry storage lockers,
and the anchor chain locker.

The next deck up was given over to the galley, a dining salon with
banquettes around a single table that could seat most of the crew of
twenty-two at the same time, a double cabin for Jacques and Simone
Cousteau, Cousteau's office, six two-berth cabins, toilets and showers,
and a wine cellar. A shelter deck ran on the port and starboard sides
from just under the wheelhouse to the open diving deck, onto which
the staterooms in the middle of the ship opened directly through separate doors. The navigation bridge, chart room, a four-berth cabin,
and the captain's cabin topped the ship, behind which was a streamlined funnel. On the funnel were painted the green silhouettes of a
swimming sea nymph and a dolphin. First sketched by painter Luc-
Marie Bayl on a bar napkin, Cousteau and Simone chose it to be the
insignia of *Calypso* and French Oceanographic Expeditions.

Cousteau made *Calypso*'s sea trials a celebration for some of the
investors who had contributed time and money during the year of
refitting. He retained a professional engineer, Octave Leandri, known
as Titi, but the rest of his crew for the voyage to Corsica was distinctly
amateur. Cousteau was the captain, Simone was in charge of stewardship, and their thirteen-year-old son Jean-Michel and eleven-year-old
Philippe signed ship's papers as cabin boys. Dumas was first mate.
Roger Gary, his brother-in-law, the Marquis de Turenne, and Pierre
Malville, who owned a restaurant in Antibes, stood wheel watches.
Jacques Ertaud, a young photographer, recorded the trip on film. All
but Titi worked without pay.

Calypso performed perfectly during the 300-mile round trip between Antibes and Calvi, the northern harbor on the island of Corsica. Cousteau learned that his ship was extremely maneuverable, able
to pivot in its own length by running one engine forward and one in
reverse. They did no scuba diving during the sea trials, but everyone
took a plunge into the warm Mediterranean off Corsica, easily
climbing back aboard using a swim step on Calypso's stern and a
short ladder to the aft deck. The weather was typical of midsummer,

warm and sunny with light winds, but it was obvious to Cousteau that his ship was sea kindly. The heavily timbered hull seemed to cling to the surface and comfortably rode the slight swells that rose in the gentle afternoon breeze. The radar and sonar, extravagances on which Cousteau had insisted because they were the most modern tools for navigation and studying the seafloor, worked fine, though learning to use them was going to take time.

As Cousteau prepared for his first expedition in the autumn of 1951, his elation about finally taking *Calypso* to sea was dashed by the news that his mother had died. Cousteau had spent very little time with her since the war. She accompanied Daniel to Sanary only a few times, choosing instead to stay closer to PAC, who was serving his life sentence for collaboration at a prison near Paris. On one of her regular trips to see him, she had a stroke that killed her. JYC, Daddy, and the rest of the family except for PAC went to St.-André-de-Cubzac and buried Elizabeth in the plot they had reserved for the Cousteau tomb years earlier. A week later, Cousteau took *Calypso* to sea on their first expedition together.

11

IL FAUT ALLER VOIR

COUSTEAU CHERISHED THE CAMARADERIE that emerged aboard *Calypso* as it steamed away from France, headed for the Red Sea. He was the commander of the expedition, but he hired former navy chief boatswain and champion offshore sailboat racer François Saout to run the ship as its captain. The rest of the hired crew were René Montupet, engineer Octave Leandri, first mate Jean Beltran from the Undersea Research Group, former navy cook Fernand Hanen, and photographers Jacques Ertaud and Jean de Wouters d'Oplinter. Dumas was aboard to supervise the diving with Cousteau. Simone Cousteau assumed the roles of chief steward and assistant to the ship's doctor, Jean-Loup Nivelleau de la Brunnière Veron.

"Many of those who came to us were sensitive young men who have not found happiness or peace leading an ordinary life," Cousteau said of his crew. "That makes them valuable to us. I cannot help thinking that the men of *Calypso* resemble, in many ways, those of Jules Verne's *Nautilus*—men who have been wounded by life on land, and who thereafter put their trust in the sea."

On the first day out of Toulon, it was obvious to Cousteau that *Calypso* was going to be a good-humored ship. Because money was tight, the ship sailed with half a full complement of professional sailors, which meant that the scientists, divers, and filmmakers had to stand wheel watches and help with maintenance and cooking. Everyone was addressed as either *Professeur* or *Docteur,* unless that person had actually earned the title, in which case they were called *Monsieur.* During the party before sailing, someone had produced a small portrait of Napoleonic general Pierre Cambronne, famous for his reply to Wellington's demand for surrender at Waterloo: *"Merde."* By acclamation, the crew declared Cambronne the godfather of *Calypso,* and hung the portrait at the head of the table in the galley. There were no distinctions of rank aboard *Calypso,* no officer's mess. Everyone ate around a single galley table that became the spiritual and social center of the ship. In his toast at dinner the night before departure, Cousteau closed with what he said had become his motto: *Il faut aller voir.* We must go and see for ourselves.

The mood aboard *Calypso* as she steamed south past Corsica was jubilant, more like a group of friends off on a cruise aboard a yacht. In keeping with French custom, everyone shook hands on first meeting each other in the morning, and again before turning in at night. On the second day Cousteau suggested that they limit the hand shaking to once in the morning. That afternoon, Cousteau encountered photographer Jacques Ertaud, who was reloading a film magazine with his hands in the black changing bag. Cousteau nodded to Ertaud, who reached up with his hands still in the bag to shake. Both men collapsed into laughter. Cousteau's greeting reform movement ended, and it was back to hand shaking morning and evening.

No one aboard *Calypso* doubted that Cousteau was their leader, but

Cousteau holds up a Roman amphora from the
waters off North Africa, 1953
(COURTESY MIT MUSEUM)

they also knew that he was always open to ideas other than his own. He seemed to function as a catalyst for innovation rather than as an innovator, far more interested in getting a job done than in doing it himself. Everyone around him recognized that his greatest talent was inspiring other people to help him realize his vision. Cousteau never surrendered authority; he was quick to criticize or dismiss harshly those who were lazy, disloyal, or incompetent; and he always seemed to know where he was going, even if he didn't say so.

Cousteau knew exactly what questions his first voyage had to answer: Is *Calypso* fit for service on the open ocean for weeks at a time? How does a private oceanographic research venture survive financially? After spending hundreds of thousands of francs refitting his ship, the French Oceanographic Expeditions bank accounts were down to zero. Cousteau had carefully chosen the scientists for the voyage who would best demonstrate that *Calypso* and its scuba divers could perform a wide variety of underwater research that would attract future funding from corporations, magazines, television networks, and scientific institutions.

Cousteau also wanted to prove to oil companies that *Calypso* and his divers could explore the seafloor for likely petroleum and mineral deposits. He invited the legendary vulcanologist Haroun Tazief and Vladimir Nesteroff, who was an expert on interpreting electronic soundings. Jean Dupas, a former French paratrooper and the third member of the geologic group, was fluent in Arabic. The hydrologic team, charged with examining the chemistry and conditions of the water itself, consisted of Dr. Claude-Francis Boeuf, Bernard Callame, and the second woman aboard, Jacqueline Zang. Marine biologist Pierre Drach, a professor at the Sorbonne; Gustave Cherbonnier, who studied mollusks; coral reef expert André Guilcher; and Claude Levy, a laboratory technician, completed the scientific party. The mission of the voyage was to locate oil, gas, and mineral resources, but Cousteau wanted scientists with him to add credibility to what he found.

On the third night out of Toulon, *Calypso* was steaming over the deepest part of the Mediterranean—the 16,500-foot Matapan Trench—when a gale roared in from the cooling deserts to the south. In an hour, the seas built to 20 feet. There was no chance for sleep, so everyone crowded into the galley and wheelhouse, keeping their spirits up together. The only thing to do was keep the bow into the wind and, at all costs, never allow the ship to fall broadside to the towering waves.

Saout and Cousteau were both in the wheelhouse when Montupet telephoned from the engine room to say that the fuel filters and lines were clogging. The first real weather *Calypso* had encountered since refitting was shaking dirt and rust loose in the fuel tanks. They would lose both engines unless he shut them down one at a time to clear the filters and line. Saout and Cousteau agreed. A minute later, they felt the first engine go silent. *Calypso* instantly became far more difficult to steer with the thrust coming from only one propeller. Cousteau had no idea how much his ship could take. He was worried that the observation tube might not be strong enough to withstand the forces of the bow rising out of the water and slamming back into it. Worse, if the second engine quit before the first was running again *Calypso* would swing sideways. How much of a broaching angle could it take and still recover? Cousteau shouted to Saout to round up as many men as he needed to rig up an emergency sea anchor. If both engines stopped, the anchor would be their only hope for holding the bow into the wind.

On the wildly pitching deck, Saout, Dumas, Beltran, and two of the scientists, Cherbonnier and Nesteroff, were struggling to build a sea anchor from a rubber life raft when the second engine stopped. At the helm on the bridge, Cousteau was helpless as *Calypso*'s bow fell off the wind and his ship slid sideways into the trough. Over the shriek of the wind and the clattering of everything on the ship that was not tied down, Cousteau screamed, "Hang on hard!" as *Calypso* rolled 45 degrees to starboard. For endless seconds, the ship lay trembling on its side with water flooding over the rail. Everyone prepared to go into the sea. Incredibly, *Calypso* shook off the water and bobbed upright. Another big wave slammed it back down, then another. Each time she recovered. Cousteau felt the blessed vibration of at least one engine starting. He swung the bow back into the wind, set a course for the lee of the island of Crete, and yelled back to Saout, "She can take it!"

After its harrowing beginning, Cousteau's first expedition to the Red Sea with *Calypso* was a thirty-eight-day triumph. The geologists found evidence of oil-bearing shale and volcanic mineral deposits using the echo sounders. Aqua-Lung divers systematically took samples from different layers of water down to 150 feet, giving the hydrologists glimpses into the complexity of seawater. Aided by divers with nets, spearguns, and explosives, the biologists caught and jugged hundreds of sea creatures, many of them new to science. Cherbonnier and Drach named three of them in honor of their ship, its captains, and the expedition commander: *Calypseus, Saouti,* and *Cousteaui.*

Over the vast coral reefs in crystalline water, Cousteau and Dumas shot color movie film underwater, and still photographer Ertaud took hundreds of pictures using Kodak Ektachrome transparency film. They used lights in waterproof housings powered by electrical cables from the surface for the deepwater work, and captured spectacular images of parrot fish, jacks, bonitos, sardines, triggerfish, and damselfish. Their pictures of corals, sponges, and other invertebrates were equally sensational, wild splashes of color that looked like organ pipes, deer antlers, and petrified plants, none of which had ever been seen by people other than pearl divers.

As *Calypso* steamed home through the Suez Canal, Cousteau and his crew enjoyed raucous, hours-long meals under the glowering portrait of Pierre Cambronne, confident that what had seemed like a pipe dream a month and a half earlier could become reality. With scientific grants and charters from oil companies and other commercial ventures, they would very likely be able to fulfill what for Cousteau was really their mission: to film the world underwater and show it to the world above.

When Cousteau got back to Toulon, he received a letter from James Dugan, the American writer who had broken the story on the first menfish two years earlier. For Dugan, the article was just the beginning. Now, he suggested that the photographs and logs from the Red Sea would make a great book. Cousteau was much more intent on eventually producing a full-length movie, but Dugan's idea made sense. No one but *Calypso* and its divers had perfected the combination of the Aqua-Lung and underwater cameras. The results would surely captivate readers as well as theater audiences. In addition to publicizing the equipment Cousteau was testing, a book might even contribute to the bottom line of French Oceanographic Expeditions.

With a suitcase full of his Red Sea photographs, his notebooks, and an outline for an article he hoped to sell to *National Geographic,* Cousteau boarded an Air France Lockheed Constellation in Paris and flew to New York to meet Dugan and Perry Miller. While he was gone, his crew would resupply *Calypso* for its next expedition to the wreck of an ancient Greek freighter reported to be off the coast near Marseille. Sunken treasure also had the potential of turning a profit for *Calypso,* and the film of Aqua-Lung divers salvaging two-thousand-year-old artifacts would be sensational.

Cousteau and Dugan decided that their book should begin with the invention of the Aqua-Lung and end with the successful expedition to the Red Sea. It would be the story of the first human beings in history to swim free underwater. Dugan happily agreed to ghostwrite the book from the notebooks and logs of Cousteau and Dumas. Dumas would be a coauthor. Philippe Tailliez had also been with Cousteau from the beginning, but he would not share credit on the cover of the book. Cousteau was still deeply attached to the memory of their time together, but when Tailliez chose the navy over *Calypso,* Cousteau's old friend became part of the past.

Leaving Dugan to prepare a proposal for selling the book to a publisher, Cousteau went to Washington, D.C., to see Gilbert Grosvenor, president, editor, and son of the founder of the National Geographic Society. Grosvenor jumped at the chance to publish the account of the expedition to the Red Sea with photographs. What else, he wanted to know, did Cousteau have planned? Cousteau was ready for the question. The society had been sponsoring expeditions since it awarded Robert Peary $1,000 for his quest to reach the North Pole, and had since funded William Beebe and Otis Barton's bathysphere descents, Auguste Piccard's balloon flights into the stratosphere, and countless other pioneering adventures.

Cousteau had no doubts that *Calypso* should fly the flag of the National Geographic Society as well as the French tricolor. He proposed an expedition to take stock of all the world's oceans using not only Aqua-Lung divers but a pressurized submarine capable of reaching depths of 1,000 feet. With *Calypso* as the mother ship for the divers and the sub, he told Grosvenor, *National Geographic* would help him banish the world's ignorance of the most vital, fascinating, and mysterious realm on the planet. Grosvenor did not sign a check on the spot, but he was clearly interested. What is your biggest problem?

Grosvenor asked. Aside from money, that would be lights for deep-ocean photography, Cousteau said. As a first step, Grosvenor suggested that Cousteau immediately meet some other members of the society's board, its staff, and a man who might help with lights. Harold E. Edgerton of the Massachusetts Institute of Technology had succeeded in freezing what the eye cannot see with strobe lights and shutter speeds of a millionth of a second, Grosvenor explained. He might be interested in helping solve the puzzles of underwater lighting.

On his way home to France, Cousteau passed through New York, where he met again with Perry Miller, who had more good news. The flood of technological innovation in cameras, film stock, and audio recording had thrown the movie business into a revolution that was bringing heightened reality and intimacy to films. One of Miller's new jobs was to find new documentaries for CBS television. The producers of a ninety-minute weekly show called *Omnibus* wanted very much to see whatever Cousteau did next.

The future for French Oceanographic Expeditions looked promising, but until *National Geographic* made its decision to support Cousteau's ambitious survey of the oceans or CBS signed a contract, he had to put *Calypso* to work to make some money. Dumas came up with what sounded like the perfect idea. A year before, one of the first independent Aqua-Lung divers, Gaston Christianini, had surfaced too quickly, got hit with a severe case of the bends, and was taken by ambulance to the Undersea Research Group's recompression chamber in Toulon. Christianini lost his toes but survived, becoming friends with Dumas, who tended him in the chamber and afterward in the hospital. He had been scraping out a living salvaging scrap from the seafloor, but after his accident he said he was finished with diving forever. Out of gratitude, Christianini spent an afternoon before leaving the hospital telling Dumas what he had seen during his hundreds of hours underwater off the Riviera. He talked about enormous piles of military scrap from which the salvaged metal was endless, secret places where there were thousands of lobsters, rock piles on the coast where he always found giant groupers and other big fish. What had interested Dumas most, however, was Christianini's recollection of piles of old jars a few hundred meters off the barren rock called Grand-Congloue, just 12 miles from Marseille.

Cousteau and Dumas knew from past experience that Christianini's jars might be amphorae from a wrecked Roman ship. They remembered the immense satisfactions of their underwater archaeological expeditions aboard *L'Elie Monnier,* and decided to risk the little remaining cash in the French Oceanographic Expeditions' treasury to retrieve at least a sample from the wreck. If it proved to be important enough for an all-out recovery job, Cousteau would approach the Borely Museum in Marseille and maybe even *National Geographic* to pay for it. With a job like that, French Oceanographic Expeditions could survive until Cousteau raised the money for his grand exploration of the world's oceans.

In midsummer 1952, *Calypso* sailed with Cousteau, Dumas, and enough men for the day trip to Grand-Congloue. The director of antiquities of Provence, Fernand Benoît, was aboard to immediately determine the provenance of the jars when they brought one of them to the surface. Cousteau anchored *Calypso* well off the jagged white cliffs of the island, and left the ship in a launch with Benoît and Dumas, who would make the first dive. Dumas went over the side wearing triple tanks and surfaced twenty minutes later. He had found the underwater arch Christianini had described but no sign of amphorae, a shipwreck, or anything that looked remotely like a jar.

Noticing Benoît's frustration at having taken a full day out of his schedule because a scrap diver had passed on a tale about sunken treasure, Cousteau said he would dive to take a look himself. Because of his trip to New York and chronic ear trouble, Cousteau hadn't been in the water in three months, but he didn't want to give up without searching a wider area in the same vicinity. Christianini had had no reason to make up the story.

Passing 50 feet, Cousteau felt good, no pain in his ears. At 170 feet, where he could clearly see the bottom, he began to feel the onset of nitrogen narcosis. No amphorae. He forced himself to pay attention, knowing that at that depth breathing ordinary compressed air he had a maximum of ten minutes left before the rapture would force him to decompress and surface.

"My right hand means south," Cousteau thought, struggling to remember how he had oriented himself during his descent. He looked south, toward the arch where Dumas had come up empty. Visibility was 100 feet. There was nothing but the gray talus and boulders of the sloping bottom.

"My left hand means north." He moved off in that direction, knowing that Dumas and the others in the launch would simply follow his bubbles to keep up with him. Fatigue reminded him that he was out of shape, and he slowed his kick as he scanned the bottom. Nothing. Then he saw something, a long dark object rising from the bottom. He descended to investigate, felt himself swooning into the fog of narcosis, and checked his depth gauge. Two hundred and fifty feet. "Stupid," Cousteau thought, kicking back up to 170 feet and encountering what he assumed to be the upward sloping edge of the island. He tripped his reserve valve to give himself an extra five minutes on the bottom.

And there it was. Looking like an object in a museum diorama of an ancient shipwreck, an amphora lay half buried on the slope in front of him. With the last measure of his strength, Cousteau pulled the amphora free of the bottom and stood it upright as a marker he or Dumas could easily find on their next dive. He ascended slowly up the slope he believed to be the rocky outcrop of the island. At a depth of 100 feet, Cousteau realized that he had been swimming over a huge mound of cargo and debris from a shipwreck. He hovered for his first decompression stop, reached into the wreckage through the fog of narcosis, and picked up what looked like three stone chalices. Ten feet from the surface, he stopped to breathe off the rest of his air, hoping it would be enough for complete decompression. His brain cleared, and he marveled at what he held in his hands.

"These cups are almost certainly Campanian ware," Benoît said, while Cousteau lay exhausted on the deck of the launch. "It is enough evidence to assume that the wreck is as old as the second century before Christ."

Dumas asked Benoît if it was worth an all-out salvage job.

"Absolutely," the antiquarian said.

If Benoît was right about the provenance of the cups, the discovery would be among the most sensational of all time in the young field of underwater archaeology, which claimed only a handful of important shipwrecks. A week later, Cousteau had promises of funding from the museum, the city of Marseille, and *National Geographic*.

Salvaging artifacts on a scale as large as what apparently lay on the bottom off Grand-Congloue was not simply a matter of sending divers down to pick things up and bring them to the surface. In the great heap Cousteau had seen there would surely be many artifacts, requiring that divers remove hundreds of tons of rock, sediment, and debris to get at them. To invent methods and machines for the specialized job, Cousteau created his second nonprofit corporation. The French Office of Undersea Technology, based in Toulon, would promote development of underwater exploration tools, patent them, and manage the revenues they generated.

For the better part of a year—with cameras rolling on deck and underwater—Cousteau and his crew worked from *Calypso* and a base on shore to raise thousands of artifacts. The amphorae, Campanian pottery, and pieces of the wreckage were from not one but two Roman ships that sank more than two thousand years ago. To clear the sediment and debris without breaking the artifacts, they used a suction dredge powered by a gas-powered compressor onshore. It was a finicky contraption, but when it worked a diver blasted air into the debris pile, loosening shards, artifacts, rocks, and wood, which were sucked into a second larger hose by the pressure differential between the surface and the depth of 130 feet. Divers also used baskets lowered by a deck crane and air bags to raise amphorae and other large pieces to the surface.

Though the salvage job off Grand-Congloue barely paid for itself, it made headlines and newsreels all over Europe. In one film clip, Cousteau is shown taking a drink of 2,200-year-old wine from one of the amphorae. The expedition became a cause célèbre on the Mediterranean coast, drawing sailboats and launches full of curious spectators to the island. Because of the publicity, the job also attracted several new crew members to *Calypso,* young men for whom the adventure was more the reward than the salary. Among them were two sixteen-year-olds, Albert Falco and Raymond Coll, and the more seasoned divers André Laban, Henri Goiran, Raymond Kientzy, Yves Girault, and Jean-Pierre Servanti. Together with Dumas and the others who remained from the Red Sea expedition, they formed what Cousteau hoped would be the nucleus of *Calypso*'s crew for years to come. They were tireless, every man brought diving and one or more other essential skills to the expedition, and Cousteau simply could not ignore the fact that they were incredibly charismatic and photogenic. He had sensational scenes of his men working underwater, but the shots of them kidding around on deck and sitting down in Roman togas for a meal on Grand-Congloue added unique dimensions of charm and character to his film. Simone was rarely seen on camera, so life as a *Calypso* diver was a man's fantasy world both above and below the surface. Cousteau was quick to recognize that Dumas, Falco, Laban, and the rest were natural movie stars.

The celebrated archaeological expedition took a tragic turn in November 1952. Because winter weather made it too dangerous to anchor *Calypso* so close to the rocky cliffs of Grand-Congloue, a team of six divers lived ashore in a base camp they called Port Calypso. On a routine supply run from Marseille, Cousteau arrived aboard *Calypso* to find that the mooring buoy to which he tied up the ship had been blown a half mile from its usual place by a storm the night before. Veteran diver Jean-Pierre Servanti volunteered to check out the situation. He discovered that the mooring chain between the anchor and the buoy had broken and the anchor was nowhere to be seen.

Other divers searched all day for the anchor with no luck. Finally, Servanti went back into the water to follow the furrow left in the bottom by the dragging chain. Though the sounder showed that the depth was 230 feet, Servanti thought he could make a quick dive and return to the surface without much decompression. Five minutes after he splashed into the water, his bubbles stopped. Falco, Ertaud, and Girault threw on tanks and were in the water in seconds, but they found Servanti lying dead on the bottom. All of France grieved with *Calypso*'s crew when accounts of the tragedy appeared in newspapers the following day.

The following spring, Cousteau tested the world's first underwater television camera on the wreck at Grand-Congloue. It was in a clumsy housing that looked like half a 55-gallon oil barrel, and sent low-resolution images through a cumbersome cable to a monitor on *Calypso,* but it worked. With it, archaeologists who were not Aqua-Lung divers stood in the galley and saw for themselves what was going on 130 feet below. They could then brief divers on what to spend their time on and what to ignore.

The video camera gave Cousteau another idea for opening the underwater world to people who could not scuba dive. At lunch one afternoon, he was thinking out loud about a highly maneuverable submarine that could do what Piccard's clumsy bathyscaphe could not do: real work underwater.

"The *commandant* took two saucers, placed one right side up on the table and the other upside down on top of it," Albert Falco remembered. " 'There: our submarine.' " When *Calypso* returned to Toulon, Cousteau sketched out the design of his diving saucer at the Undersea Research Group workshop. He wanted a two-man submarine that could be launched from *Calypso,* reach a depth of at least 1,000 feet, and move as freely through the water as a scuba diver.

12

FAME

WHEN COUSTEAU COMMISSIONED James Dugan to write the story of the invention of the Aqua-Lung and the wonders of the undersea paradise he had opened to the world, he knew the book would be an important historical document. He had no idea that it would be wildly popular. In a little over a year, Dugan had written *The Silent World: A Story of Undersea Discovery and Adventure, by the First Men to Swim at Record Depths with the Freedom of Fish.* A month after publication in February 1953, *The Silent World* was on the *New York Times* best-seller list, where it remained through the summer. The first half of the book is an account of the invention of the Aqua-Lung by Cousteau and Gagnan, followed by riveting tales of *Les Mousquemers* clearing mines, cave diving, shipwreck diving, and their encounters with fish, dolphins, seals, and sharks.

Rachel Carson reviewed *The Silent World* for the *Times.* Her own 1951 book, *The Sea Around Us,* proposed that the oceans were not indestructible but fragile natural treasures threatened by the growing human population. It had won the National Book Award and been on the best-seller list for eighty-six weeks. What Carson wrote in her review of *The Silent World* anointed Cousteau as her ally and a powerful force for transforming the human relationship with the sea:

> Beyond its ability to stir our imagination and hold us fascinated, this is an important book. As Captain Cousteau points out, in the future we must look to the sea, more and more, for food, minerals, petroleum. The Aqua-Lung is one vital step in the development of means to explore and utilize the sea's resources.

By the end of 1953, *The Silent World* had sold 486,000 copies and was being translated into French and twenty other languages. The

Jacques Cousteau (in wet suit), *Louis Malle* (right),
and an unidentified man on Calypso
with underwater television apparatus
(COURTESY MIT MUSEUM)

success of the book improved Cousteau's financial picture, but the royalties weren't enough to finance an expedition to turn *The Silent World* into a movie. Every bit of his cash was going to meet payrolls and expenses for the venture at Grand-Congloue and his Office of Undersea Technology. He was expecting a grant from the French Ministry of Education, but the collapse of the government forced him to start over with new bureaucrats in Paris. *National Geographic* continued to encourage him, but had not written any checks. In December 1953, just as Cousteau was beginning to think he might have to fold the research center and tie up *Calypso,* he got lucky.

On a wet afternoon, while Cousteau was ashore in Marseille, a man decked out in a banker's suit appeared at *Calypso*'s gangway and asked permission to come aboard. Simone showed him into the galley, offered him a whiskey, and asked him what brought him to the waterfront on so dreary a winter day. He said he represented the

D'Arcy Exploration Company, a subsidiary of British Petroleum, and he had a proposition. Simone said she knew of British Petroleum since her cousin, Basil Jackson, was president of it. The man said the chief of D'Arcy Exploration had read *The Silent World* and thought that Aqua-Lung divers might help his company prospect for oil. Would Captain Cousteau consider a four-month charter of *Calypso* and its divers in the Persian Gulf? An hour later, Cousteau returned and agreed to a fee that was nowhere near the cost of an ocean oil exploration rig but was more than enough to save him. The D'Arcy Exploration had no objections to having a film crew aboard.

By the spring of 1954, everyone in France seemed to know about the bold and handsome Aqua-Lung divers who were retrieving stunning artifacts from ancient Roman wrecks and making underwater motion pictures of their adventures. Cousteau received a steady stream of letters and inquiries from men and women who wanted to join in the adventure, regardless of the danger, the long voyages away from home, and the meager pay. He was looking for sailors, engineers, and divers. He also wanted moviemakers. To scout for talent, Cousteau went to a film festival at the Institute of Advanced Cinematographic Studies in Paris. There, he met Louis Malle, a compact, sharp-featured young Frenchman born to wealth, who was a recent graduate of the institute. Malle was looking for a job.

Louis Malle was the fifth of seven children of Pierre Malle and Françoise Beghin, the heir to a sugar beet fortune built by her ancestors during the Napoleonic Wars. His mother went to mass every day of her life, and insisted that her son's education be steered by the Catholic church, first at French boarding schools and then with the Jesuits in England. The family summered in Ireland, where Louis mastered English and settled into himself as a citizen not only of Europe but of a wider world revealed to him in theaters. The flowering of French cinema threw blossom after magnificent blossom in front of him. Jacques Tati, Jean Renoir, and Robert Bresson were directing dazzling films on the aftermath of the war, love, revenge, and the rest of the human condition. Malle could not resist making a wholehearted attempt to join them. His parents had plotted a far more conservative course for their fifth child, which would have taken him into the management of their sugar beet empire, but they continued to support him after he turned his back on business in favor of cinema.

Cousteau could see himself in Malle. He recalled that his own young life had been a similarly uneven quilt of absent parents, new surroundings, a succession of different schools, and the irresistible lure of moviemaking: Daniel Cousteau's constant traveling with his American employers; the sojourn to the United States, where Jacques had become comfortable with the English-speaking world; boarding school after the rock-throwing incident; and then his fascination with movie cameras, which saved him during adolescence.

Like Cousteau, Louis Malle had stories to tell. In 1944, with the Allies advancing on occupied Paris, his father had predicted a bloody battle for the city and sent Louis to a monastery near Fontainebleau, where he would be safer. The Carmelite monks, whose role in the resistance would become legendary, were sheltering as many Jewish children as they could feed until they were betrayed by a kitchen worker. With twelve-year-old Louis Malle watching in horror, retreating Germans seized the Jewish boys, none of whom would survive the war. Years later, Malle would make two films on the participation of the French collaborators in the horror of the Third Reich, branding himself as a controversial filmmaker. After coming of age in occupied France, Malle knew that evil and good were present in every human being. Only the passion of the moment really mattered. On the evening he met Cousteau in Paris, Malle could not have expressed his cinematic vision so clearly, but later he was to say, "Each movie is a piece of life, a different adventure. It expresses my interest of the moment, somewhat like a love."

On their first meeting, Cousteau asked Malle what kind of films he wanted to make. Malle said he believed that both documentaries and dramatic features revealed human events and passion, so he was willing to try anything. He had seen Cousteau's film *Épaves* several times and thought that underwater films were somehow between fact and fiction because they revealed so alien a world. Cousteau made his offer. He had just published a book about the invention of the Aqua-Lung that was selling well in the United States and wanted to make a movie from the same material and additional film from a two-year expedition back to the Red Sea and into the Indian Ocean. Cousteau told Malle that he had very little money, so he could pay almost nothing. Malle said the money wasn't important.

Cousteau used part of the advance payment from British Petroleum to buy a 60-foot fishing boat to finish the job at Grand-Congloue, freeing *Calypso* to move on to another adventure. The day after Christmas 1954, he called Louis Malle in Paris and told him to catch the next train to Marseille. A week after New Year's, with the Cousteaus, Dumas, Malle, Laban, Falco, twelve other crewmen, and a Portuguese water dog named Bonnard aboard, *Calypso* sailed again for the Suez Canal. After nine months of repetitive industrial diving on the Roman shipwrecks, everyone was overjoyed by the prospect of a few months on the open ocean.

In the Red Sea, Louis Malle became a qualified Aqua-Lung diver and proved to be an inspired choice as head cameraman. He immediately began improving the cameras, lights, and housings, and, best of all, he shared the enthusiasm of everyone else for the wonders he saw below the surface. Underwater, Malle, Cousteau, and Dumas anticipated each other's moves as if they had been diving together for years. On deck, Malle moved easily among the crew with an instinct for the quirks, humor, and personality that might charm an audience. Malle was completely comfortable in the cramped little observation chamber on *Calypso*'s bow and filmed an enormous school of dolphins from it off the coast of Yemen.

Cousteau took a month rounding the Arabian Peninsula into the Persian Gulf. In Elphinstone Inlet, a narrow fjord between limestone cliffs in the Strait of Hormuz that is reputed to be the hottest place on earth, they found delicious oysters that no one knew existed. Farther up in the gulf, Malle filmed the remains of the once-legendary pearl diving industry that had been displaced by cultured pearls from Japan. The pearl divers were all old men who dove with no goggles, fins, or snorkels, reminding Cousteau and Dumas of their earliest days together as free divers on the Riviera. On the bottom, at 60 feet, Malle's camera captured them groping like blind men but somehow coming up with full baskets of oysters.

As Cousteau and *Calypso* turned north into the Persian Gulf, the petroleum geologists briefed the crew on the history of oil. Since the first wells began pumping oil in Titusville, Pennsylvania, in 1859, petroleum exploration had spread around the world, first to banish the night with kerosene lamps, then to power the machines and automobiles on which the modern world depended. In 1922, after Royal Dutch Shell had been drilling successfully in the basin surrounding Lake Maracaibo in Venezuela for ten years, Standard Oil of New Jersey took the risk of drilling wells beneath the surface of the lake. Some Standard Oil executives thought their venture was folly, joking that they would be better off going into the fishing business, but the Lake Maracaibo field turned out to be one of the most productive in history. In 1932, a British-American joint venture had struck oil on the little island of Bahrain in the Persian Gulf. It was 20 miles off the Arabian Peninsula and made of rocks that were geologically identical to those on the already proven oil fields on the mainland.

With absolutely no restrictions on subsea oil exploration or production, the rush was on. Afterward, the constantly improving technology of offshore drilling opened the continental shelves of the oceans to petroleum production. Hard-hat divers could explore the bottom looking for rocks and sediment that might indicate the presence of subsurface oil, plant the legs of drilling rigs, and cap producing wells with pumps. But they moved painfully slowly, and they were limited to depths under 200 feet. The Aqua-Lung, British Petroleum suspected, could change everything.

During the next three months, *Calypso* made four hundred stops to prospect for oil. At each of them, they swung a large bell-shaped machine called a gravimeter over the side with the crane mounted on the aft deck. On the bottom, the gravimeter measured fluctuations in the force of gravity, certain types of which indicated there was oil-bearing rock and sediment below. *Calypso* divers then descended to retrieve samples from the bottom, which in the beginning seemed impossible because of the hardness of the rock almost everywhere in the gulf. The petroleum geologists, who had come aboard at Aden on the trip around the Arabian Peninsula, were well aware of the density of the Persian Gulf seafloor because they had dulled or broken countless drilling bits during the past decade. When Dumas and Falco made the first dives, they tried to extract rock samples with chisels and sledgehammers. They got their samples, but swinging the hammers through the dense water exhausted them and drastically cut their bottom time because of hyperventilation. Dumas thought one of the ship's paint chippers powered by an air compressor on the surface might work better, but when he tried it, each burst of the chipper comically bounced him ten feet off the bottom. They went back to the hammer and chisel. What had begun as a high-spirited adventure became a tedious series of challenges every bit as difficult as the salvage job off Grand-Congloue.

When *Calypso* anchored overnight off a desert island just after entering the gulf, Cousteau, Dumas, and several others had gone ashore, where they found a single man living in an air-conditioned hut and tending a navigational radio transmitter. The long-haired hermit who answered the door gasped, pointed at Dumas, and said, "You're Frédéric Dumas!" By sheer coincidence, he had been reading *The Silent World* when the *Calypso* divers arrived at his desolate outpost. That was cause for celebration, so they took the man, who introduced himself as Tony Mould, back to *Calypso* for a good meal. Before dinner, Laban gave him a haircut while everyone took turns quizzing Mould about the snakes and sharks they had heard infested the Persian Gulf. Mould told them that he never went into the water himself, but the local pearl divers had been burying their dead on his island for years. There were twenty-two graves, Mould told them. Two dead from sharks, the rest from sea snakes.

The twin scourges of the Persian Gulf presented Cousteau and his crew with an impossible trade-off, which became the subject of spirited discussions at every meal. *Calypso* carried a steel cage to which the divers could retreat from menacing sharks, but once inside they would have little maneuvering room to dodge the snakes, which seemed to appear in swarms. All the divers had seen sharks underwater—never so many and never so aggressive a species as the blue sharks that infested the gulf, but at least they were familiar creatures. The sea snakes were something else. Called golden snakes by *Calypso* divers, they have venom that attacks the nervous system, like the venom of the krait, a deadly southern Asian snake. Death from respiratory collapse or heart attack comes within minutes of a bite. There was no known antidote. Supposedly, even a 6-foot snake has a mouth so small that it can bite only a small fold of skin, such as the tendon between the thumb and forefinger, but that was little comfort to the divers.

During three months in the Persian Gulf, *Calypso* divers took twice as many samples from the bottom as their contract required. The oil geologists marked the locations of promising sites on their charts, triangulated by radio signals from navigational transmitters like the one run by Tony Mould. D'Arcy Exploration and British Petroleum were so happy with the work of Cousteau and his divers that they promised much more for the future and named the 12,000 square miles they had explored the Calypso Grounds.

With the petroleum geologists off the ship, and three months of exhausting industrial diving behind them, *Calypso* and its crew sailed for home, planning to take a month to explore and film whatever caught their attention. When they stopped at Doha, Qatar, on their way south out of the gulf, Cousteau received a telegram from Daddy Daniel. The French Ministry of National Education had finally awarded a grant to French Oceanographic Expeditions. In exchange for two-thirds of FOE's annual budget, *Calypso* would carry scientists to be named by the National Center for Scientific Research on missions to coincide with Cousteau's own plans for exploration and filming. *Calypso* had become the official French national oceanographic research ship. The first scientists, Daddy wrote, would be a three-man team of marine biologists who would join *Calypso* in Doha the following day for the voyage back to Marseille.

With James Dugan, who had flown south to do research, and the three biologists aboard, Cousteau sailed south through the Strait of Hormuz, the Gulf of Oman, and into the Indian Ocean, where he took a detour for a week along the coast of Africa. The scientific mission was vague, but the biologists seemed pleased with whatever happened each day. Everyone was relieved to be free of the grueling diving schedule of oil exploration, and much of what they encountered they were seeing for the first time: flying fish, sea turtles, a scarlet raft of eggs that stretched as far as they could see. The scientists dissected and jugged what the divers brought back, Louis Malle was always in the right place at the right time, and Cousteau reveled in the future while at the same time being riveted in the present.

Calypso, its crew exhausted after six months at sea, returned to Marseille for repairs. Nothing major had gone wrong with the ship, but Cousteau's next expedition, financed by the new grant from the government, was going to last more than a year. He overhauled both engines and caulked seams in the hull that had sprung during a storm in the Mediterranean, during which the ship's dog, Bonnard, had been lost overboard.

When Cousteau returned to France, he learned that his brother, Pierre-Antoine, had been granted a mercy parole from prison because he had terminal cancer. Soon after PAC's release, his wife, Fernande, died, also from cancer. Their daughter, Françoise, was old enough to be living on her own, but though Jacques and PAC were still on the chilliest of personal terms, PAC's son, Jean-Pierre, spent summers and vacations in Sanary-sur-Mer with Daddy Daniel or aboard *Calypso* if the ship wasn't too far away. The rest of the year, he attended boarding school in Normandy with the Cousteaus' sons, Philippe and Jean-Michel.

"I didn't much care for the schools my mother and father sent us to," Jean-Michel said. "What I lived for was our time on *Calypso* with them. We ate dinner together every night. We were together all day. We dove, worked on equipment, and saw parts of the world most boys only dream about. It was an intense, wonderful time, what they call quality time nowadays."

In the workshop at the Office of Undersea Technology, Cousteau, Dumas, and Malle rebuilt their cameras with new lenses, drives, and housings that had been developed while they were at sea. They were adaptations of standard Bell and Howell 35 mm movie cameras, which they called Sous Marine (Underwater) or SM One, Two, and Three. They also improved on their still cameras, adding strobes and floodlights similar to those they used with the movie cameras.

On the Persian Gulf expedition, Cousteau, Dumas, and Malle had learned that their biggest problem in filming in deeper water was bringing enough light with them. The year before, while *Calypso* was between supply trips to Grand-Congloue, Harold Edgerton spent a month on the Riviera testing remotely controlled lights and cameras for photographing in depths down to 1,000 feet. Edgerton was a loquacious, endlessly curious man who had broken free of the restrictions of academic life with his work on stroboscopic lighting, which famously allowed him to photograph the impact of a drop of milk and a bullet piercing a playing card. Edgerton was also a pioneer in the development of the side-scan sonar, which was part of *Calypso*'s electronic inventory. At MIT he was a beloved character whose philosophy of teaching, he said, was "to teach people in such a way that they don't realize they are learning until it is too late." When Edgerton and Cousteau had met for the first time in New York two years earlier, they had agreed that the key to education lay in fascination. Edgerton was happy to help Cousteau light his underwater movies, and became a regular visitor to *Calypso,* where the crew dubbed him Papa Flash. His son Robert, who came to France as his assistant, was Petit Flash.

Calypso, bright white with a new coat of paint, sailed from Marseille in early March 1955 on a four-month expedition to explore the ocean from the Red Sea to the Seychelles Islands in the Indian Ocean, and down to the northern tip of Madagascar off the coast of Africa. Biologists and geologists from the National Center for Scientific Research joined the ship at various points on its voyage, but the overarching mission of the expedition was to shoot enough film to assemble the movie version of *The Silent World.* With Cousteau in command, Simone—*La Bergère*—tended the details of life aboard ship. Dumas was in charge of diving. Louis Malle was chief cameraman with Papa Flash's lights. No one among the crew of twenty-five had the slightest doubt that they would succeed. At the last minute, *National Geographic* came through with a grant in return for exclusive rights to an illustrated story on the expedition under Cousteau's byline, the star power of which increased steadily as sales of *The Silent World* soared over a million copies. The *Geographic* also sent photographer Luis Marden, who was willing to learn to dive with an Aqua-Lung as well as record the expedition topside.

The now familiar trip south on the Mediterranean and through the Suez Canal was one long celebration. *Calypso* was a more genial ship than ever—constant hand shaking, basking in the sun on deck, card playing, and the loud, endless meals. With twenty-five people and Simone's dachshund, Bulle, aboard, quarters were tight, but there wasn't a ripple of irritation, even in heavy seas. Cousteau put into Port Said on the Mediterranean, then Port Sudan on the Red Sea for fuel, water, and supplies, but except for three days investigating the wreck of a British ship sunk by German planes during the war, he kept moving. He had plenty of footage of the Red Sea from two previous expeditions, and he had high hopes for what he would find in the pristine, unexplored coral reefs off Assumption Island in the Seychelles.

Four hundred miles off the coast of Kenya, a day from Assumption Island, *Calypso* steamed into a pod of sperm whales. Several adults with young swam in random patterns around the ship, loping through the water as though in no great hurry to get where they were going. Cousteau throttled back, but not before *Calypso*'s bow slammed into one of the whales. It was seriously wounded, with blood streaming in its wake. Cousteau and his crew stood in stunned silence as two other whales swam to the injured whale and supported its body with their own. In minutes, all the other whales had converged on the scene of the accident, swimming slowly in a defensive perimeter around the wounded member of the pod. With *Calypso* barely moving to keep steerage, Cousteau felt one of his propellers hit something. Seconds later, a 15-foot-long infant whale broke the surface next to the ship with deep, bleeding gashes clearly visible on its back.

As though by a single command, the pod and the first wounded whale slowed and fell behind *Calypso,* vanishing completely just as the first sharks began tearing at the crippled infant. Dumas ran to the weapons locker on the navigation bridge, grabbed a rifle, and killed the little whale with a single shot. While he was administering the coup de grâce, Cousteau, Laban, and Malle rigged the shark cage to the crane. In minutes, they were in the water capturing a terrifying feeding frenzy never before seen on film. Malle was shaking so badly he had to struggle to hold the camera steady as he watched 12-foot blue sharks methodically rip chunks of flesh from the whale, then follow the skeleton into the depths to scavenge the last morsels.

Afterward, as always, Cousteau's evening included updating the ship's log, in which he recorded his amazement over the behavior of the sperm whales as much more profound than the horror of the sharks' feeding frenzy. During the encounter, he had listened to the sounds of the whales with *Calypso*'s echo sounder, hearing the unmistakable cries of distress from the whale wounded by *Calypso*'s bow. Cousteau told everyone that he believed the whales had spoken to one another, both during the collision and just before the shark attack, when the entire pod vanished, as if on command. He had never seen or heard anything like it, and knew that his film from that afternoon would change the way the world thought about whales and sharks. The shark frenzy had disturbed the bucolic rhythms of life aboard *Calypso*. Before going back to work, Cousteau gave his crew a week to unwind in Victoria, Mahé, the northernmost port in the Seychelles.

When they finally reached Assumption Island, a tiny chunk of upthrust limestone fringed with a pristine coral reef north of Madagascar, even the most seasoned divers were stunned by the colorful natural masterpiece below. As always, Dumas and Falco made the first reconnaissance dive on the reef. They had seen more beneath the sea than any two men alive, but both returned to the surface sputtering through their mouthpieces, *"Extraordinaire."* At dinner that night, Cousteau announced that *Calypso* would stay at Assumption Island until the monsoons arrived in about a month. The coral reef and its creatures, Cousteau thought, were the most marvelous movie set he could imagine.

Cousteau promoted Malle from head cameraman to codirector. They had spent hundreds of hours together planning their dives, repairing cameras, lights, and hydrophones for sound, and talking about their expectations for *The Silent World*. To Malle, Cousteau was a master technician who constantly strove to improve his equipment and technique for filming underwater. To Cousteau, Malle was the embodiment of the artistic sensibility he considered essential to great cinema. At twenty-three, Malle already believed that film—like dance, theater, and music—allowed him to marry beautiful images, great writing, interesting characters, and music into a story that happened, ended, but remained forever in the consciousness of the viewer. After fifteen years of underwater film, it was no longer enough, Cousteau and Malle agreed, to simply take people beneath the sea. Cousteau wanted a pure documentary, a movie that revealed not only the underwater world and the creatures that inhabited it but the divers responsible for making it. Malle wanted to portray the divers as dancing rather than laboring, making them not so much guides and masters underwater but strange humans weirdly adapted to the alien world. They agreed that *The Silent World* would be shot in color.

They compromised on the kind of film they would make together, trading scenes in a carefully edited dance of pictures and sound. Malle's opening mesmerized its first audiences at the Cannes Film Festival in the spring of 1956. Five Aqua-Lung divers descend through aquamarine water carrying underwater flares that stream pyrotechnic banners of bubbles, accompanied by the sounds of their breath through gurgling regulators and the roar of the burning flares.

"This is a motion-picture studio sixty-five feet under the sea," Cousteau narrates, his voice conveying fascination in every word. "These divers wearing compressed-air Aqua-Lungs are true spacemen, swimming freely as fish."

The descent seems to last forever. People in theaters find themselves holding their breath. A hundred feet down, the flares die out and the divers turn on floodlights that illuminate a riot of orange, yellow, and red bursting from a coral reef beneath them. They change course and continue to descend. Passing 200 feet, the ocean is a diffuse bluish haze that collapses into blackness below. The *Calypso* divers are now in the world of rapture, Cousteau continues, with a note of fear in his voice. Nitrogen in their flesh and blood will soon intoxicate them, robbing them of their sense of balance and their ability to make decisions. A minute later, the divers are at 247 feet, the deepest Aqua-Lung dive ever captured on film. In the weird white beams of the camera lights, the divers nod to each other, point to the surface, and begin their ascent.

The second scene is Cousteau's, a distinct counterpoint to Malle's lyrical overture. The divers are back on the surface, exhausted and struggling to board *Calypso* with the weight of their tanks and the awkwardness of walking in swim fins. Cousteau's obvious point is that menfish swim free in the sea but are as clumsy as fish out of water on the surface. The first lines of wooden dialogue reveal that great divers are not necessarily great actors, but somehow they strike just the right note. One of the divers complains of a pain in his knee. It might be the bends. Cousteau orders the diver into *Calypso*'s recompression chamber, where he will have to stay while the rest of the crew eats dinner.

"Do I have to, Captain?" the diver protests, sounding like a scolded boy.

"Absolutely," Cousteau says, pointing to the chamber as if banishing that child.

After a quick tour of the interior of the chamber and an explanation of the bends, laughing crewmen seal the pouting diver inside. Cousteau's technical moment segues into the galley, where the crew,

crammed around the impossibly small table, feasts on lobsters and
talks about diving. All of them are awkward on camera, but they are
infinitely likable and unquestionably brave. Everyone in the theater is
quite willing to dive with them again.

For the next eighty minutes, Cousteau, his divers, and a cast of sea
creatures make good on the dramatic promises of the opening scenes.
A 60-pound giant grouper named Jo Jo le Merou (after a famous
Marseille gangster) dances a waltz with Dumas, who feeds him scraps
of meat from his hands. Cousteau explains that the fish off Assumption Island have no fear of humans because they have never seen any
before. He also says that he has banned spearfishing by anyone except
the *Calypso*'s chef. Jo Jo is shown greeting the divers in the morning
and following them to the boarding ladder in the evening. The big
fish stuck so close to the divers that they had to lock it in the shark
cage when they filmed other reef fish, lest it interrupt the scene by
chasing the smaller fish.

Cousteau leads his divers and theater audiences to the wreck of a
sunken freighter in the Red Sea with footage shot on his first and second expeditions. Sixty feet down, he films Dumas scrubbing rust and
silt from the ship's builder's plate, revealing that it was from a Scottish
shipyard, yard number 599. Back aboard *Calypso,* Cousteau scrapes
barnacles and algae from a bell retrieved from the wreck, showing the
camera that the ship's name was *Thistlegorm.*

In another of Louis Malle's scenes, hundreds of porpoises cavort in
front of *Calypso,* easily able to outswim the ship but clearly slowing
down to play, apparently happy to entertain Cousteau and his laughing
crew standing on the ship's bow. Not a single member of the audience
at Cannes or in the thousands of theaters in which *The Silent World* was
shown had ever seen anything like the dance of the dolphins.

Nor had they ever seen the brutal antithesis of the dolphins' grace
and beauty: the shark feeding frenzy on the baby sperm whale, shown
while Cousteau matter-of-factly comments on the attack. "The sharks
smell the blood in the water. Then comes the first bite. It is the signal
for the orgy to begin." *Calypso*'s crewmen, Cousteau explains, are outraged by what the sharks are doing. He calls the sharks the mortal enemies of divers. The crewmen bait hooks, catch the sharks, then join the
orgy themselves by hacking them to pieces on deck. At the screening
of *The Silent World* in Cannes, audiences moaned and gasped at the carnage on the screen. Many wept.

During the festival, *Calypso* was moored in the bay, its rigging strung with lights and pennants, its white hull an unmistakable nautical centerpiece. For the first time, the jury awarded its highest prize, the Palme d'Or, to a documentary. A week later, after making an initial approach to Daniel Cousteau, Columbia Pictures bought the rights to release *The Silent World* in the United States.

Four months later, *The Silent World* opened at the Paris Theater, across the street from the Plaza Hotel in New York City. The other-worldly divers in the opening scene had people holding their breath, Jo Jo le Merou (renamed Ulysses in James Dugan's English narration) had them laughing, the dolphins were more beautiful than anything they had ever seen on the ocean, and audiences were similiarly caught up in the shark frenzy. Bosley Crowther of the *New York Times* called it "the most beautiful and fascinating documentary of its sort ever filmed. The only trouble with the whole thing is that it makes you want to strap on an Aqua-Lung and go."

The following spring, after playing to enthusiastic audiences in hundreds of theaters across America, *The Silent World* won the Academy of Motion Picture Arts and Sciences Oscar as the best documentary film of 1956. Crowther and other critics, however, wondered about the scientific value of running *Calypso* into a pod of whales, slaughtering sharks for revenge, or dynamiting a reef to collect fish. "Exactly what Captain Cousteau learned for the benefit of oceanographic science is not explained," Crowther wrote. "However, his voyaging turned up a beautiful and absorbing nature film, and that is enough for anybody whose scientific interest does not range very far outside a theater."

Among those who were enchanted by the exploits of Cousteau and the *Calypso* divers was Prince Rainier of Monaco. After seeing *The Silent World,* he offered Cousteau a job as the director of the world's oldest undersea museum and research center, the Oceanographic Museum of Monte Carlo. The museum, which looks like a gigantic limestone castle hanging on the face of a 500-foot cliff outside the harbor, was built by Rainier's grandfather, Prince Albert Grimaldi. Calling it the Temple of the Sea, he filled it with a collection of specimens and artifacts from his own ocean expeditions. The museum was equipped with laboratories, meeting rooms, and an extensive library, which, for the rest of Albert's life, drew the cream of European ocean scientists to Monaco.

The harbor at Monaco (COURTESY OF THE AUTHOR)

Monaco itself is less than a square mile of land, bordered on three sides by France and on the fourth by the Mediterranean. Until Louis Blanc, a gambler exiled from Germany, arrived in 1872, the principality was a few narrow streets winding over a precipitous rock cliff, a fishing fleet of a dozen small boats, and a population of eight hundred people who scratched out a wretched existence under the guard of a battalion of French troops. With the roulette wheel he brought with him from Bad Homburg, Blanc transformed Monte Carlo into the gambling capital of the world in less than a decade. He ensured his welcome and the continuing health of Monaco by cutting in the Grimaldi family, heirs to the throne of the principality, for 10 percent of his action, which quickly amounted to millions of francs a year.

Albert's heir, Prince Louis, inherited the throne in 1922. He had no interest in the sea or its creatures, so the museum went into a quarter century of decline. When Prince Rainier took over in 1949, he made the resurrection of his grandfather's vision one of the priorities of his reign, spending the next eight years rebuilding the now decrepit fortress on the cliff. In 1957, the museum was just beginning to attract not only tourists but scientists again. Rainier decided that Jacques-Yves Cousteau would be the perfect man to raise it up the next notch to its former glory as one of the world's great centers of inquiry into the nature of the world's oceans.

Rainier saw Cousteau as a celebrated explorer who also knew how to tell the world what he saw underwater in books and movies. He was a master fund-raiser and a great showman, traits that fit perfectly into Rainier's dream that the museum would, like the casino at Monte Carlo, become a source of revenue for the tiny principality as well as contributing to scientific knowledge of the ocean. The prince envisioned it becoming a self-sustaining aquarium, featuring creatures from the Mediterranean, in particular the dolphins that had so moved him in *The Silent World*.

Cousteau told the prince that exhibiting live dolphins in an aquarium had been a dream of his since his first expeditions aboard *Calypso*. He found out that American marine parks captured dolphins by lassoing them, and put his research group to work figuring out how to do it. They mounted a platform on the front of one of their launches, where Falco stood with a rope and a long pole to place the noose over the animal riding the bow wave. It didn't work. Falco killed a few dolphins before giving up. Cousteau then realized that the kind of dolphin the Americans captured with lassos was the bottle-nosed dolphin, a much more robust animal than the common dolphin in the Mediterranean, which was lighter and more delicate. Falco tried anesthetizing the dolphins with curare before lassoing them. More dolphins died.

Finally, Falco got a line on a small female, let her tire herself out pulling against a buoy on the surface, and sent divers into the water to corral her. They brought the exhausted dolphin to shore in Marseille, put her into a large concrete tank, and named her Kiki. She died three months later, but not before bonding with Falco, who was her primary keeper. Coincidentally, or maybe not, Kiki died a week after Falco went to sea aboard *Calypso*, leaving her with a new keeper. Cousteau insisted that she died of a broken heart when Falco left. Two more dolphins, a male and female, killed themselves by swimming full speed into the wall of the tank. With much better facilities and the help of scientists at the Oceanographic Museum, Cousteau hoped he would eventually succeed in holding dolphins in captivity and make their leaps and playful swimming a major attraction in Monaco.

Rainier's offer was a dream come true for Cousteau, who constantly was struggling to make financial ends meet, both for his expeditions and his family. It included a generous salary, an apartment on a hillside in Monaco, and a staff to handle the day-to-day operations of the museum. He was free to manage his movie production company, *Calypso,* the Office of Undersea Technology, and his role as the spokesman for Air Liquide's scuba equipment venture. Like Bosley Crowther, almost everyone who saw *The Silent World* couldn't wait to go see the world beneath the sea for themselves. Aqua-Lung sales were booming in the United States and Europe.

With the security and status of his appointment to the museum in Monaco and the success of his other enterprises, Cousteau no longer harbored the illusion that he would ever return to service as a full-time naval officer. His sons were nearly grown. Jean-Michel was studying to become an architect after finishing his two years of military service. Philippe, two years younger, was still sampling his future, most of which revolved around the passions of his father—flying, engineering, diving, and cinematography. Cousteau's decision to become a civilian, however, was not an easy one. He had spent twenty-seven of his forty-seven years in the navy. As an officer on inactive duty, he had maintained his uniforms, received a small monthly paycheck, and had the use of navy facilities and equipment when he needed them. He was, however, the lowest-ranking member of the naval academy class of 1933 as a lieutenant commander. The decision was difficult for Simone as well. Though she shunned the tight-knit culture of officers' spouses in favor of life aboard *Calypso,* she was the daughter and granddaughter of French admirals. Simone had never experienced life without the navy. It was clear to both of them, however, that the navy no longer needed the Cousteaus and the Cousteaus no longer needed the navy. Shortly after accepting Prince Rainier's offer, he resigned his commission.

13

LIVING UNDERWATER

DURING THE SUMMER OF 1958, Cousteau's first after returning to civilian life, he tested the revolutionary research submarine on which he and the Undersea Research Group had been working for five years. Its chief engineers, André Laban and Jean Mollard, were finally ready to launch the saucer-shaped craft they called simply Hull Number One. On its cradle aboard *Calypso,* the yellow metal submarine stood 5 feet high with a diameter of 6 feet 7 inches, looking very much like the two saucers Cousteau had clapped together on the table to show Falco and the rest of his crew what the shape of a nimble research submersible should look like. Since October 1948, when *Les Mousquemers* had accompanied Auguste Piccard on an expedition to test his deep-diving bathyscaphe, Cousteau had been tantalized by the ocean beyond his reach with scuba gear. The bathyscaphe had been perfectly suited for setting depth records and gathering samples from a small area of the bottom, but it was far too clumsy for real undersea exploration and filmmaking. Hull Number One could carry a crew of two, who entered through a hatch in the top of the saucer, lay on their bellies, looked through a pair of Plexiglas viewing ports, and steered with buttons that controlled swiveling jet thrusters.

"Cousteau really did come up with the idea that it should look like a saucer at lunch that day off Grand-Congloue in the early fifties," André Laban recalled. "But it took a lot of people a lot of time to figure out the details."

Laban was in charge of the first unmanned descent to a depth of 2,000 feet at the end of a cable. If Hull Number One survived, he would then send it to 3,000 feet before approving it for manned dives to no more than 1,000 feet. Laban and Mollard had modified an industrial crane on *Calypso* that could lift 5 tons, more than the

La Souscoupe (AGENCE FRANCE-PRESSE)

weight of the submarine and ballast for the unmanned test, and the much lighter weight of the manned sub and its steel cradle.

Off the coast southeast of Marseille, Laban lowered Hull Number One to a depth of 2,000 feet. He let the craft soak for fifteen minutes, during which a mistral kicked up a swell from the south. When the sub was 100 feet from the surface, clearly visible to the anxious crew on *Calypso*'s pitching deck, the cable snapped, almost beheading the winch man as its end lashed back. Falco dove into the sea but was helpless as Hull Number One plunged 3,300 feet to the bottom.

Cousteau quickly took radar bearings on three shore targets to mark the spot. The following day, Laban returned aboard *Calypso* and conducted a sonar survey of the bottom for 10 square miles around the site. He, Mollard, and the other engineers at the Office of Undersea Technology were already working on Hull Number Two from the same design. If Number One had not survived the descent to 3,000 feet, they were wasting their time. When Cousteau and Laban analyzed the sonargrams, they clearly saw that the lost sub was hovering over the bottom, secured by a cable they knew to be 30 feet long that tied it to the 4 tons of lead ballast required for the test descent. Obviously, it was still airtight at 3,270 feet. They could go ahead on Hull

Number Two without the expense of retrieving the first one. Cousteau knew it was only a matter of time before he added 750 feet to his reach into the sea.

The following year, 1959, after what had seemed to have been an endless streak of good luck for Cousteau, PAC died. He had battled cancer for five years while continuing to publish unapologetic accounts of his war years as a collaborator. His estrangement from his brother had deepened, as JYC increasingly kept his distance to avoid a public connection with a man who had betrayed his country. A few hours before PAC lapsed into the coma from which he would not wake, JYC was at his bedside to offer his once-beloved brother something more than his friendship in life. For the rest of his own life, JYC would care for PAC's children as his own. Cousteau never intended to separate himself from PAC, never condemned him as a traitor. He simply excluded the man who had been his best friend until politics and war took him away, learning from that painful detachment that all relationships are transitory. Cousteau could be immediately and passionately present with people, but he also wore a hardened suit of emotional armor that allowed him to move on without them.

Seven months after PAC's death, Cousteau sailed for New York to attend the World Oceanographic Congress at the United Nations. It was the first international gathering of its kind, with a thousand scientists and explorers from thirty-eight countries convening across the ordinarily impassable borders of academic disciplines to assess the future of the world's oceans. Cousteau was one of the featured speakers, and the mayor of New York promised him and *Calypso* the kind of welcome usually reserved for the maiden voyages of great ocean liners and heroic men-of-war.

Crossing the Atlantic was no mean feat for the 139-foot, flat-bottomed *Calypso*. Cousteau and the crew knew their little ship could take a pounding and stay afloat, but the ride was anything but comfortable. Most days, the rule was one-hand-on-the-ship to keep their footing, and even the veteran sailors among them were not immune to seasickness during the worst of it. North of the Canary Islands, *Calypso*'s crew and a team of French oceanographers under the leadership of American geologist Lloyd Breslau took historic photographs of the mid-Atlantic rift zone using cameras towed on underwater sleds and Papa Flash's new lights. The sleds, nicknamed Troikas after horse-drawn Russian sleighs, took pairs of color photographs that could later be viewed stereoscopically, giving depth and dimension to the bottom terrain. The mid-Atlantic rift, a great gash in the seafloor discovered during the *Challenger* expedition in 1863, runs 18,000 miles from Greenland to the fringe of Antarctica, marking the line where the supercontinent of Pangaea tore apart 180 million years ago and started the formation of the current landmasses of the earth. The theory of plate tectonics, spreading seafloor rift zones, and drifting continents was new, controversial, and the most thrilling discovery in the history of the young science of oceanography. Geologists, biologists, physicists, chemists, and metallurgists were crossing interdisciplinary lines in droves to contribute to this new idea about how the planet was formed and to explain the remarkable role of the deep ocean in creating the earth's crust. They had sampled the midocean ridges and rifts with dredges and probes, and sketched their contours with echo sounders, but actually getting a look at big black cliffs in photographs was fantastic.

The mayor of New York made good on his promise. Cousteau and *Calypso* passed the Statue of Liberty under fireboat water showers, accompanied by sirens, horns, and whistles. Overhead, a dirigible hovered with a camera crew filming the brave little ship. Her white hull was weathered and streaked with rust from the long Atlantic crossing, her crew at the rails waving to their escorts through the fog of the August morning. For the next week, *Calypso* was open to the public at her berth on the Lower West Side, while across town Cousteau attended the congress and Breslau's presentation of the spectacular photographs of the mid-Atlantic rift. In dozens of sessions, oceanographers outlined a revolutionary new human relationship with the world's oceans. Increased knowledge and the rapid development of technology to exploit ocean resources were transforming the industrial world. It was also obvious that the sea was not the infinite source of wealth and life it had been thought to be just a decade earlier.

Cousteau had recently seen firsthand how quickly human impact can transform a piece of the resilient, beautiful, fruitful ocean into a wasteland. His first big idea as the director of the Oceanographic

Cousteau and Simone aboard Calypso *in New York during the*
World Oceanographic Congress in 1959
(ASSOCIATED PRESS)

Museum had been the creation of a marine sanctuary beneath the
cliffs of Monaco that would be tended by Aqua-Lung divers and
viewed by visitors ashore via television cameras. He called it the
Marine Biotron. In it, divers and scientists would live in underwater
houses to study a pristine section of the ocean that did not experience
the pressures of fishing, sport diving, and most of all coastal development. Cousteau was forced to abandon the Biotron when towns on
both sides of the proposed preserve, Fontvieille and Monaco Beach,
embarked on aggressive landfill projects to expand their territory in
the tiny principality, with no limits on the amount of concrete, rocks,
dirt, and gravel they poured into the sea. For months, the water was
clouded with sediment that killed the delicate marine organisms upon
which fish and other sea life depended. Worse, Cousteau learned, the
breakneck expansion of nuclear energy plants in southern Europe

meant that the entire Mediterranean coast was going to be polluted with low-level radioactive waste. Cousteau had been exploring and filming the underwater world as much for the sheer pleasure of the adventure as anything else. After the World Oceanographic Congress, there was no question in his mind that by showing as many people as possible the beauty, power, and paradoxical vulnerability of the world's oceans he could help save them from destruction.

Cousteau and *Calypso* left New York to show the flag on the Atlantic seaboard, sailing first to the Woods Hole Oceanographic Institution on Nantucket Sound, then south to the Potomac River and Washington, D.C. There, as the guest of honor at the National Geographic Society, Cousteau was presented with a gold medal and his crew were treated as visiting dignitaries. A never-ending stream of people who considered themselves to be friends of Cousteau's great adventure toured *Calypso*. A writer and photographer from *Time* magazine came aboard to interview Cousteau for a brief article, which grew into the lead feature story.

On the cover of *Time*, Cousteau's beaming, weathered face appeared at the center of a montage depicting a scuba diver, reef fish, and corals. In the upper-right-hand corner, a banner announced: "SKINDIVING. Poetry, Pleasure, and Pelf." Inside, the story began with Cousteau's description of the transformation he and every scuba diver experience underwater: "From birth, man carries the weight of gravity on his shoulders. He is bolted to earth, but man has only to sink beneath the surface and he is free. Buoyed by water, he can fly in any direction—up, down, sideways—by merely flipping his hand. Underwater, man becomes an archangel."

The story filled ten pages, with photographs, covering Cousteau's life, the early adventures of *Les Mousquemers,* the invention of the Aqua-Lung, *Calypso* and its charismatic crew, *The Silent World,* the revolutionary new and untested submarine in *Calypso*'s cargo hold, and Cousteau's recent appointment as the head the world's oldest oceanographic museum in Monaco. There were more than a million Aqua-Lung divers in North America, all of whom had bought their equipment from U.S. Divers, the Air Liquide subsidiary. Cousteau was its president. The writer concluded by asking Cousteau what advice he would give to a person trying scuba diving for the first time.

"What would you advise a baby to do when it is first born?"

Cousteau replied. "When a person takes his first dive, he is born to another world."

Calypso's crew had enjoyed the festivities surrounding their port calls in New York, Woods Hole, and Washington, but everyone was happy to get back to the routines of life at sea. They savored the hand shaking, tinkering with equipment, mealtime conversation, and even the nicknames. Cousteau was Pasha, because he was the oldest man. Albert Falco was Bébert. Dumas was Didi. The crowds swarming aboard *Calypso* and requesting autographs had been heady, but most important, their fame meant that there were no limits on what they could propose and accomplish as filmmakers and explorers. Cousteau and his divers had proved their worth on so many fronts of oceanography and the popularization of the sea that they could now choose from an unlimited pool of adventures and make money doing it. They headed south to finally test their new submarine.

Hull Number Two, renamed *La Souscoupe Plongeante*—The Diving Saucer—because of its resemblance to a comic book flying saucer, was an improved version of the lost prototype. Twin propulsion jets on the bow of the flattened sphere swiveled on command from the pilot, giving it an infinite range of motion. On a strut that extended from the starboard bow—determined on the round saucer by the location of two viewing ports—was an Edgerton stroboscopic camera, with its synchronized light mounted on a similar extension from the port bow along with the movie camera floodlight. The camera itself was inside the sub, so the crew could reload it on long dives. It was mounted on a bracket between the two viewing ports and trained through its own window. Three small Plexiglas ports in the top of the dome gave pilot and observer a view above, and three sonar transducers up, down, and above transmitted their signals to a screen on the instrument panel. The pilot and observer lay on foam cushions surrounded by gauges showing air pressure, oil pressure, sonar readings, depth, voltage in the battery-powered electrical system, the amount of carbon dioxide and oxygen in the atmosphere, and a compass. The oxygen rebreathing system and carbon dioxide scrubbers could keep two people alive for twenty-four hours. If the sub was crippled, the pilot could pull a hand lever releasing ballast to send it to the surface.

Laban and Mollard had worked frantically but failed to finish the sub before *Calypso*'s scheduled cruise to New York for the World Oceanographic Congress in the summer of 1960. On the voyage across the Atlantic and at the dock in New York, they finally got Hull Number Two ready for testing.

Cousteau decided to test the second diving saucer in just 80 feet of water on the Caribbean continental shelf off Puerto Rico. If anything went wrong, there would be no problem retrieving the sub. The first manned descent off Puerto Rico, with Falco and Mollard aboard and a winch cable still attached, lasted only fifteen minutes. The second, to a depth of 100 feet, lasted an hour, after which Falco popped from the hatch shouting, "What a hot rod!" He had been able to control the descent of the neutrally buoyant saucer by pumping mercury ballast forward to point the bow down, level off by neutralizing the ballast, and spin on the saucer's axis with the propulsion jets. Visibility through the observation ports was excellent. He had been able to follow a swimming grouper almost as effortlessly as a scuba diver. The noise of the motors and pumps startled fish, but the presence of the big yellow saucer didn't seem to drive them completely away.

For the third dive, they removed the safety cable and went to 200 feet. Cousteau remained on deck for the first eight dives, knowing that if the saucer got into trouble his decisions would be critical. Next, Harold Edgerton, who had joined *Calypso* at Woods Hole, became the first scientist to dive in *La Souscoupe*. In his diary, Papa Flash recorded the clarity of the view through the Plexiglas ports and the clean taste of the air he was breathing. "Being in the saucer is no different from being in an automobile," he wrote, "except that we are more comfortable and loll on our mattresses like Romans at a banquet."

The next day, with Cousteau and Jacques Ertaud outside in scuba gear to film the diving saucer maneuvering, Falco took the sub through its paces. Writing on a white dinner plate with a grease pencil, Cousteau set up the shots. "Spin," he wrote, holding the plate up to Falco's porthole, then backing away with the camera. Falco spun the saucer. "Lazy eights," he wrote, asking Falco to perform a well-known airplane maneuver, then watched the saucer fly through water as easily as a plane through air. They went on for an hour, returned to the surface for fresh tanks, and went back down for one last shot of the saucer dropping the emergency ascent ballast. Cousteau wrote, "Drop in one minute," pushed the plate up to the porthole, and was starting to back into position for the shot when he heard and felt the hollow thud of an underwater explosion. He darted back to the window. The inside of the saucer was completely dark. For long seconds, the Plexiglas was a mirror reflecting only Cousteau's own face. Finally Falco's replaced it, his puzzled expression letting Cousteau know he had no idea what had happened. Cousteau left the window, looked around the hull of the saucer, and saw streams of bubbles coming from the fairing on which the camera lights were mounted. He wrote "Battery Fire" on the plate and thrust it at the view port. Falco's face instantly disappeared. Cousteau heard the clatter of the ascent ballast release and the sub shot to the surface.

In its cradle on Calypso's deck, *La Souscoupe* spewed billows of smoke. Falco and Mollard popped the hatch and scrambled out. Laban and Edgerton tore open the fairings around the battery compartment and hit the smoldering fire with CO_2 extinguishers. The nickel-cadmium batteries were brand-new, supposedly a vast improvement over heavier, less powerful lead batteries, but they had never been tested underwater. Cousteau and Laban hoisted the saucer and dropped it back into the sea with the battery cases open, finally extinguishing the fire. During the postmortem, Cousteau discovered that his new batteries had not shorted out and caught fire. They had given off so much heat that the fiberglass boxes filled with oil that surrounded them had caught fire.

Three weeks later, after dissecting the batteries to find nothing wrong with them and refitting the sub with brass battery boxes that vented the heat, Cousteau and Falco prepared to dive into Inferno Bay in the Cape Verde Islands. They lowered the saucer on the end of a cable, fully powered up, to a depth of 1,500 feet with no problems at all. The batteries in their new housings worked perfectly.

Cousteau had been living in anticipation of his first dive in *La Souscoupe* for seven years. He made two dry runs on deck. When he and Falco settled into their cushions, the launching was second nature to both of them. Falco closed the hatch. Cousteau set the pressure on the oxygen valves. Falco uncovered two racks of CO_2 absorbent and flicked a switch to start the hydraulic motors controlling the steering system. Cousteau gave the thumbs-up for launch to Laban, who was replacing him as dive master on deck. The interior flooded with the rippling aquamarine light of the shallows. Outside, scuba divers gave them an underwater acrobatic show. *La Souscoupe* sank 100 feet and came to rest on a sloping, sandy bottom. Falco pumped mercury ballast into the forward tank, pointing the sub's nose downslope, and squeezed a burst from the propulsion jets. Hovering at 4 feet, they descended along the bottom as it fell away into the depths. Two hundred feet . . . 250. "From now on we're on our own," Cousteau said. "The divers can't help us if anything goes wrong."

At 300 feet, the temperature in the sub plunged suddenly as it passed through a thermocline separating distinct layers of water. Cousteau and Falco put on sweaters. At 360 feet, they felt the sub scrape bottom, even though Falco had not put it there. For some reason, *La Souscoupe* had lost positive buoyancy. Falco shut down the motor, hoping that the silence would give them a clue about the condition of the sub. Then they heard it, a teakettle hissing of bubbles from the area around the battery boxes. Falco grabbed the lever to release the ascent ballast. The sub began to rise through a dense cloud of bubbles that should not have been there. On the instrument panel, the battery voltmeter pegged on zero. They definitely had a short circuit, and the batteries were venting gas. Through the view port, Cousteau and Falco saw the plankton moving up instead of down. They were sinking again. Falco seized a knife from its case on the wall and slashed at the tape securing the handle of the lever to release a 450-pound emergency weight from the bottom of the sub. Jettisoning the weight was the last resort, never before used. It worked. The nose of the sub tilted up 35 degrees and they were on their way to the surface. During the fifteen minutes it took *La Souscoupe* to reach the world of sunlight, Cousteau and Falco ate chicken sandwiches and shared part of a bottle of Bandol. Cousteau had brought the wine, made at a vineyard near his home in Sanary-sur-Mer, in anticipation of celebrating a successful dive to 1,000 feet. He and Falco agreed that being alive was reason enough for a toast.

Two months later off Corsica, equipped with the old, heavy, lead batteries that did not catch on fire, *La Souscoupe* took Cousteau and Falco to 1,000 feet and returned without incident to the surface. During the next year, with Falco or Laban at the controls, the diving saucer made fifteen dives in the Mediterranean Sea, carrying geologists and biologists to places none of their kind had ever been before. With *La Souscoupe*'s hydraulic claw, they collected samples of mud, rocks, sediment, and dozens of sea creatures, many of them new to science. Falco and Laban kept logs on tape and filmed their descents with movie cameras, while Aqua-Lung divers captured the nimble little yellow submarine descending into the darkness of the abyss.

After their successes on the mid-Atlantic rift, in New York, and with *La Souscoupe,* Cousteau and his divers were more capable of visiting the world beneath the surface of the sea than any other human beings in history. Now, Cousteau declared when *Calypso* returned to Marseille at the end of 1960, they would find a way to live there for extended periods of time. It was not a new idea. In the seventeenth century, the freethinking bishop John Wilkins advocated the development of underwater houses to colonize the oceans. Two hundred years later, American Simon Lake built wheeled submarines with hatches through which his submariners could leave and return while submerged. Recently, Commander George Bond had been trying to persuade the U.S. Navy to explore the possibility that scuba divers could work for days and even weeks underwater by living in gas-filled shelters on the bottom instead of surfacing. Bond called it "saturation diving," which meant that a diver's blood and tissues became saturated with nitrogen, which caused no ill effects as long as the diver didn't return to the surface, where decompression sickness could injure or kill him. Saturated divers could do all kinds of work that hard-hat divers and ordinary scuba divers could not do, such as installing offshore oil rigs and pipelines in deep water and setting up antisubmarine sonar networks. The U.S. Navy wasn't interested. Bond, who had been awed by *The Silent World,* called Cousteau and asked him to help test his radical concept for working underwater.

"I have long felt that undersea exploration is not an end in itself, although it is spiritually rewarding merely to be an onlooker," Cousteau said. "The privilege of our era, to enter this great unknown medium, must produce greater knowledge of the oceans and lead to assessment and exploitation of their natural resources. In the end man must and shall colonize the deeper ocean floor."

The first phase, Continental Shelf I—known as Conshelf I—was

simple. Two divers would live for one week in a watertight steel cylinder 18 feet long and 8 feet in diameter, anchored with chains 7 feet
above the bottom at a depth of 37 feet. The divers would work several
hours each day to depths of 80 feet, but never shallower than the
depth of the cylinder. They would enter and leave their home
through an open hole in the bottom of the cylinder called a moonpool, the water kept out by the air pressure inside.

Since successfully testing *La Souscoupe,* Falco had been Cousteau's
de facto second in command aboard *Calypso.* He led the Conshelf
expedition, christening the cylinder *Diogenes* after the beggar-
philosopher who lived in a bathtub on an Athens street. Cousteau
selected another *Calypso* diver, Claude Wesly, to join Falco in becoming the world's first aquanauts. Wesly, who was thirty years old, had
been with *Calypso* since the first expedition to the Red Sea. He had a
parrot that was older than he was, a raucous, foul-mouthed bird that
usually lived ashore with Wesly's wife and daughter but had made
short trips aboard *Calypso.* Wesly pleaded to take his parrot with him.
The bird could easily make the quick descent to *Diogenes* in a pressure
cooker and would, like a mineshaft canary, die first to warn the men
if the air became toxic. Cousteau and Falco said no.

On the morning of September 14, 1962, *Calypso* towed *Diogenes*
from Marseille to a bay on Frioul, an abandoned island that had last
been home to a yellow fever quarantine hospital. *Espadon,* a power
barge packed with equipment, friends, families, reporters, and photographers, followed *Calypso.* In a little more than an hour, scuba
divers had set the anchors to hold *Diogenes* in place, turned on the
remote television cameras inside and outside of the cylinder, and
checked the connections of tubes from the surface carrying air, water,
and electricity. A few minutes after noon, Falco, a bachelor, said
goodbye to his mother, and Wesly kissed his wife and daughter and
stroked the parrot, which they had brought with them for the day,
and they were gone.

Falco and Wesly had worked on the Conshelf cylinder from design
to outfitting. There were no surprises when they surfaced in the open
moonpool in its floor and looked around at the little room that would
be their home for a week. Somehow the fact that it was underwater
made it thrilling. Though facilities were Spartan, everything they
needed was there. Berths stood 2 feet apart on one end, with shelves

over them holding a radio, books, a television set that could receive the French national broadcasting station, and a picture of *Calypso*'s bottom from below painted by André Laban, who, for ten years, had been perfecting techniques for actually painting underwater. Across the moonpool on the other end of the cylinder were a table and chairs, a counter for serving the food that would be prepared on *Calypso* and sent down in pressure cookers, and a hot plate to reheat it if they didn't eat it immediately. Four electric fans, two on each end, circulated the air pumped from the surface. In the center, over to the moonpool, were racks for their Aqua-Lungs, emergency air tanks, fins, masks, cameras, and other equipment. Next to the equipment racks was a shower with an unlimited supply of hot water piped from above. The ocean outside would be their toilet, eliminating the need for that kind of plumbing.

Five hours after they arrived in their underwater home, Falco and Wesly were saturated with nitrogen and would have to undergo reoxygenation of their bodies to return to the surface. If the cylinder ruptured or a fire broke out, or in some other life-threatening emergency, they would bail out with their Aqua-Lungs and wait for help from above. Raymond Kientzy, who was the alternate oceanaut for the Conshelf I mission, was in command of fifteen scuba divers who were standing by in shifts in full equipment twenty-four hours a day on *Calypso*. Two doctors swam down twice a day to conduct two-and-a-half-hour examinations of the aquanauts, including electrocardiograms and blood tests.

For the first two days, Falco and Wesly were like boys in the most exotic playground they could imagine. They greeted visiting divers in *Diogenes* with harmonica duets, performed little skits for the television cameras that were always on, and frolicked outside for the ten-man film crews that spent more than an hour each day below. Falco and Wesly wore light blue gloves to distinguish them from the other divers, and seemed to swim with a touch of arrogance, knowing that everyone but they were limited to a couple of hours, at most, on the bottom.

When the doctors popped through the moonpool on the third morning, there was no harmonica duet to greet them. The aquanauts were subdued, no longer mugging for the cameras or asking for details about life on the surface. For the first time, neither of them asked about their families. Falco said he hadn't dreamed in years but was having nightmares about a disembodied hand strangling him. He was also plagued by constant visions of the air pumps failing and the lower pressure in the cylinder allowing it to completely flood through the moonpool. Wesly was sleeping okay but having trouble dealing with the moods of the once indefatigable Falco.

On the fourth day, Cousteau sent down a psychologist to evaluate the crew of *Diogenes*. He administered a battery of psych-technical and motor-function tests, on which Falco and Wesly were within normal limits. Except for Falco's anxiety and Wesly's growing irritation, they were doing fine, considering that they were cooped up in a small space and forbidden from returning to the surface. What would help, the psychologist asked? Fewer doctors, Falco told him. Cousteau reduced the daily physicals to one. What else, he asked? Send us a phonograph and some classical records, Wesly replied.

The exhaustion of coping with their living environment and a steady schedule of chores outside *Diogenes* was the real problem, the doctor told Cousteau that night. That and the fact that they were living in a small, single room. The next undersea habitat would have to contain more than one room. Further invading their privacy was the constant intrusion of phone calls from the surface day and night. Cousteau ordered a ban on all but essential phone calls. From then on, Falco and Wesly were left alone except for food shipments, doctor's visits, and the camera crews filming them when they were working outside *Diogenes*. The television monitors inside showed them usually doing nothing, listening to music, lying in their beds bundled up in wool pullovers, fleece-lined boots, and the red knit watch caps that had been favored by hard-hat divers for decades.

Under the new regime of increased isolation from the surface, Falco and Wesly improved and got some work done. They built a net corral for fish, sampled a bed of shrimp, and explored an ancient shipwreck that no one knew was there until they found it on one of their meanderings around the bottom. On the sixth day, their last, Falco and Wesly entertained Cousteau for lunch. He brought down a tin of caviar and a bottle of wine, which he had trouble opening because of the air pressure in the cylinder. Falco noticed Cousteau having trouble with the cork and asked his commandant to whistle a tune. Cousteau pursed his lips and blew, but no sound came out. Falco and

Wesly collapsed in laughter, then whistled together. Everything is different here, even learning to whistle, Wesly said.

On their last night, after the celebratory lunch with Cousteau, Falco predicted in his diary the future of living underwater. "The Pasha is thinking of deeper stations," Falco wrote, "several buildings constructed in stages—a Himalaya in reverse with Base Camp One, Camp Two, and so forth on down where we would stay weeks, even months to work. The Pasha is eloquent, full of ideas—the wine or the pressure? He talks about colonizing the continental shelf. We would all live underwater with wives and children. We would have schools and cafes. A real Wild West! I can see Claude as Sheriff of the Deep."

On the morning of the seventh day, Falco and Wesly lay on their cots wearing face masks to breathe a mixture of 80 percent oxygen and 20 percent nitrogen, approximately the reverse of ordinary air. The usual way to return to the surface would have been to recompress in a chamber for many hours until their blood gases had returned to normal, but Cousteau and his dive table expert Jean Alinat were certain that the high-oxygen mixture would do the job with far less risk and discomfort. It worked. Falco and Wesly surfaced just after one in the afternoon, with no ill effects.

"The sun is good," Falco said as he stood on shaking legs aboard *Calypso*. "The land is beautiful."

"What would you like?" Cousteau asked him.

"To walk," Falco replied.

14

WORLD WITHOUT SUN

A MONTH AFTER *Calypso* towed *Diogenes* back to Marseille, Cousteau delivered a speech at the World Congress on Underwater Activities in London. The implications of the Conshelf I success for underwater oil and mineral development were enormous, Cousteau said. The divers so necessary for anchoring and maintaining pumps, pipelines, and drilling rigs on the bottom were no longer constrained by their own human physiology. They could work underwater for days, even weeks, as long as they did not come to the surface. Cousteau pointed out that they could remain at depth in a compression chamber as well as in a habitat on the ocean floor. Conshelf I had proved the concept. Cousteau moved on to his main point, a recitation of what he had told Falco and Wesly over caviar and wine on their last day aboard *Diogenes.*

"A new species of human being is evolving," he said. *"Homo aquaticus."*

Murmurs and shuffling among the audience of oceanographers, geologists, biologists, and divers broke the silence in the lecture hall. Cousteau pressed on.

"After living in compressed air habitats for a generation, Water People will be born at the bottom of the sea. They will breathe by extracting oxygen directly from water after operations to surgically implant gills in their throats, bringing humanity full circle back into the sea," Cousteau told his startled audience of ocean scientists. "Diving has gone beyond sport. It is now a worldwide movement. The imperative need now is to place swimmers underwater for very long periods, to really learn about the sea. I think there will be a conscious evolution of *Homo aquaticus,* spurred by human intelligence rather than the slow blind natural adaptation of species. We are now moving

*Cousteau, with microphone, greets André Laban emerging from
the Conshelf III sphere in 1965* (ASSOCIATED PRESS)

toward an alteration of human anatomy to give man almost unlimited
freedom underwater."

For fifteen uncomfortable minutes, the delegates listened as a man
whose credentials as an undersea explorer were unassailable demol-
ished his reputation for responsible scientific inquiry. Water people
would be born at the bottom of the sea by the year 2000, he predicted. They would have lungs filled with an incompressible liquid,
and breathe by a blood-regenerating technique, and swim to depths
of a mile. Beyond that, he said, the pressure of 170 atmospheres
would crush a human body.

Several delegates told the press that *Homo aquaticus* was pure science fiction and wondered if Cousteau had lost his marbles. Cousteau
responded by questioning what was wrong with science fiction as a
predictor of future reality. Look at Jules Verne, he insisted. His visions
of submarines and men swimming free underwater had come true.

"The informed human imagination has projected what was to
come for hundreds of years. Actually, I was trying to be conservative

in talking about the underwater future in London," Cousteau said. "Why, there are people at the congress who wanted to talk about milking whales in regular underwater dairies. We know practically nothing about the depths of the ocean."

Regardless of his flight of fancy in London, Cousteau found plenty of interest in continuing his undersea habitat experiments among oil exploration companies. He was also sure the adventures would make another great movie. Cousteau announced that he would accomplish three goals. First, a team of divers would spend an entire month living in a base station 33 feet, or 2 atmospheres, down and working at depths to 60 feet. Second, another team would spend a week living at 82 feet, or 3.5 atmospheres, and working at depths down to 160 feet. Finally, the divers at the upper station would assemble and install an undersea garage for *La Souscoupe* to maintain and use the diving saucer without ever returning to the surface.

Calypso's schedule for the remainder of 1962 included missions to test Papa Flash's invention of a seismic transceiver that could pick out hard objects buried several feet in sediment and to sample the chemistry of seawater around the Mediterranean. Cousteau remained ashore to seal a deal with the French petroleum consortium to partially finance the next step in realizing his prediction that humans would eventually return to the sea. The oilmen were in for about half the $1.2 million Cousteau needed for Conshelf II.

The proposal Cousteau and James Dugan made to Columbia Pictures at the end of 1962 was an irresistible blend of *Calypso* divers, the exotic reefs of the Red Sea, and the tension created by putting seven men on the bottom of the ocean to see if they would survive. After the success of the far less ambitious *Silent World,* the movie studio pushed the other half of the Conshelf II budget into the pot. By the new year, Cousteau had all the money he needed.

In February 1963, Cousteau again took *Calypso* through the Suez Canal on a reconnaissance voyage to find the right location for his undersea village. After three hard weeks of diving every day in a new location, he found the perfect spot 27 miles north of Port Sudan, about halfway down on the western side of the Red Sea. The city had an international airport for resupplying the expeditions and a deepwater port, and, best of all, it lay at the heart of one of the greatest coral reefs on earth. Cousteau had a host of technical challenges to overcome in his quest to live underwater, but he was very aware that the world beneath the sea had to look good to theater audiences. The necklace of coral off Sudan was also perfect for him, because it dropped into the abyss in steplike tiers. Cousteau could anchor both his shallow- and deep-water stations to a flat bottom, which would make everything, especially handling *La Souscoupe,* much easier than on a gradually sloping reef.

"If we could make a station on the Sudanese reef," Cousteau declared at a press conference he called to launch Conshelf II, "floor settlements would be possible in other remote and inhospitable climes where the sea had hoarded her bounties."

Three months later, in June, Cousteau was back at Port Sudan with *Calypso* and the chartered Italian cargo ship *Rosaldo.* The freighter carried 500 tons of prefabricated steel for the main station, the diving saucer garage, and the deep station; lead ballast, cables, hoses, food, wine, water, and air compressors; and tanks of oxygen, nitrogen, and helium. While *Calypso* shuttled to and from Port Sudan with supplies and visitors, *Rosaldo* remained moored as the permanent surface support base. The main underwater station, named Starfish House because of its shape, had five rooms in four arms around a central hub, with an open moonpool on the bottom. The garage, anchored 50 feet away, was a steel hemisphere pressurized to keep out the water into which *La Souscoupe* could be guided by divers through a large hole in the bottom.

Falco led the Starfish House team, with Wesly, ship's cook Pierre Guilbert, and marine biologists Raymond Vaissière and Pierre Vanoni. Their quarters were air-conditioned; the food and wine were equal to that in a good restaurant in Marseille; and they had music, television, telephones, sunlamps, games, and large windows through which they could watch fish, working divers, and the diving saucer when they were off duty. They could even smoke—which all of them did— because they breathed ordinary air pumped down from the surface. This time, Cousteau let Wesly bring his parrot, so they had a pet. They submitted to daily physicals and were constantly monitored by television cameras in every room, but Starfish House offered the precious gift of privacy, without which Falco and Wesly had fallen into moodiness and depression during Conshelf I.

Starfish House was mainly a movie set and a test of endurance, but

Deep Cabin, the two-man station at 82 feet, was a far riskier exploration of the limits of the human body. To counteract the buildup of
nitrogen at 3.5 atmospheres and deeper dives to 400 feet, the Deep
Cabin aquanauts breathed a mixture of helium and oxygen. After
two abortive attempts by a dozen divers to anchor the one-room cylinder that would be their home, ten-year *Calypso* veteran Raymond
Kientzy—called Canoe because he had once fallen out of one—and
André Portelatine moved in. Within hours, it was apparent that they
were in for a miserable week. The temperature, which all had
assumed would be relatively comfortable at that depth, was over 90
degrees, worse even than that on the surface, where their support
crews at least caught a hint of a breeze. Kientzy and Portelatine were
prohibited from rising above the depth of Deep Cabin, so they were
also denied the occasional visits to air-conditioned Starfish House,
which were prize plums for camera crews and divers from the surface.

During the second week, Simone Cousteau dove to Starfish House
to escape the brutal heat aboard *Rosaldo* and spent the afternoon helping Guilbert make dinner, after which she refused to return to the
surface. She called up for clothes and toiletries and took one of the
three visitor's berths for the night. Back on the surface, after a careful
ascent through recompression stages, *La Bergère* announced that the
celebration of the twenty-sixth anniversary of her marriage to Captain Cousteau would be held at Starfish House on July 12.

Two days before the end of Conshelf II, the Cousteaus toasted each
other with flat champagne, its gas bubbles remaining suspended in liquid under two atmospheres of pressure. A procession of divers, cameramen, and a pair of oil company geologists swam down from the
surface to celebrate with them. Their younger son popped through
the moonpool to wish them well. Philippe, twenty-two years old and
finished with military service, had flown south to join the expedition,
fitting smoothly into the mélange of skills and personalities aboard
Calypso. Their older son, Jean-Michel, intent on a life on his own as
an architect, had remained in Paris.

"Simone and I had passed many unusual wedding anniversaries,
but this was the most imaginative and symbolic," Cousteau wrote
later. "*Calypso*'s festive good humor was carried to our house on the
ocean floor."

The party in Starfish House also celebrated the end of major photography for the documentary Cousteau had promised to Columbia Pictures. Louis Malle was gone, making movies on land that were already fixing his star in the firmament of the great French film-makers. Pierre Goupil, who had been Malle's assistant cameraman, had taken over as chief of *Calypso*'s film crew. The ingredients of *World Without Sun* were about the same as those for *The Silent World*. The plot was simple: Put men on the bottom of the ocean. Place obstacles in the way of their survival. Show them overcoming those obstacles. Bring them back safely. The contribution of a vast store of scientific knowledge about what happens to human beings under those conditions was the real payoff of the expedition, but the entertainment was in the details: *La Souscoupe* descending 1,000 feet into the abyss and returning like the family sedan to its bright yellow garage; men playing chess and listening to Mozart while 35 feet underwater; the sweating, obviously distressed occupants of Deep Cabin, who sounded like Donald Duck when they spoke because the helium they were breathing caused their vocal cords to vibrate faster than they do in ordinary air. The *Calypso* divers were again menfish in an alien world, confidently swimming among sharks, into and out of underwater buildings, and playing with fish that seemed to have become their pets. Two of them adopted a vicious-looking barracuda they named Jules, encouraging the fish to follow them around by tapping on their Aqua-Lung tanks. Guilbert, the cook, had conditioned a triggerfish to swim to the moonpool to receive food scraps when he tapped on a window. The reef itself was a character, blooming at night into spectacular displays of bioluminescence as its creatures ventured forth or revealed themselves under cover of darkness.

World Without Sun opened in theaters just before Christmas 1964. Audiences were huge, the film won Cousteau his second Oscar for best documentary, and critics, for the most part, applauded. A few skeptics, including Bosley Crowther of the *New York Times,* accused Cousteau of faking some of the most dramatic scenes in the movie. How, Crowther wondered in his review on December 23, could *La Souscoupe* have really entered a deep-sea cavern more than 1,000 feet down in the last scene of the movie? Does such a cavern exist? If it did, wouldn't it be filled with noxious gas instead of the air that allowed Falco to open the hatch and look around? Obviously, Crowther pointed out, the camera had to have been set up outside in the cavern to film Falco opening the hatch. How did that happen? In another scene, Crowther said, the camera moves from inside Starfish House, through a window, and out into the darkness of the sea at night. "It is too bad that this obvious faking should finally excite one's doubt and mar one's complete enjoyment of this otherwise plausible film," Crowther wrote.

Cousteau was livid. In a letter that took up three columns in the *Times* three weeks later, he wrote that the film never says the cavern is 1,000 feet down. In the dialogue between him and Falco, Cousteau pointed out that he says, "We're going up" before the scene in the cave. As for the tracking shot out the window, Cousteau said, it was a matched interior and exterior sequence that had taken an entire day to get right. Crowther called Cousteau in Paris after he read the letter, and Cousteau admitted to him that the entrance to the cavern was really only a few fathoms beneath the surface and not airtight.

Crowther had the last word. "My original objection to the staging of this incident is not altered," he wrote in the *New York Times*. "It still tends to deceive, and provoked some gnawing skepticism as to the validity of the rest of the film. This is too bad, because *World Without Sun* is so exciting it doesn't need a tricky kicker at the end."

Regardless of the objections of Crowther and others that *World Without Sun* depicted fantasy as well as fact and therefore should not be called a documentary, theater tickets sold steadily well into 1965. For Cousteau, the share of the gate was a windfall, but not enough to keep his expanding ocean enterprises running in the black. He had two hundred scientists, engineers, divers, administrators, and sailors on the payroll now. To run his Office of Undersea Technology, *Calypso,* and a fleet that had swelled to a dozen ships and launches, Cousteau had to come up with at least $2 million a year. His business philosophy placed no absolute value on money. His fiscal strategy was simply to go out and get more money when he started to run out.

With Conshelf I and II, Cousteau had demonstrated that men could live and work for weeks at a time at depths down to 100 feet. Now, the French petroleum engineers told him, show us how deep a diver can go to work on a mock-up wellhead. Offshore drilling in the Gulf of Mexico, the Persian Gulf, and off the coast of California was proving the existence of vast pools of oil beneath the seafloor. To fully exploit it, divers had to find a way to work at depths of 300 feet and more.

On September 17, 1965, *Calypso* towed a yellow-and-black 20-foot steel sphere from the dock at Monte Carlo. Inside, six men were already at 11 atmospheres after staged pressurization that took place at the dock. During the first two Conshelf expeditions, divers could reach the aquanauts in minutes if something went wrong. The deep sphere, however, would be moored 325 feet down off Cape Ferrat on the coast just east of Monaco. For twenty-seven days, the divers of Conshelf III would be as isolated from the world above as they would have been in space. André Laban was in command, with Oceanographic Museum physicist Jacques Roillet assigned to medically monitor the crew, the working divers Raymond Coll, Yves Omer, and Christian Bownia. Philippe Cousteau, who had just graduated from the French national film school, was the cameraman.

Before *Calypso* and the Conshelf sphere reached Cape Ferrat, a mistral blew up, forcing the ship and the sphere it was towing back to the harbor at Monte Carlo to wait out the storm. The men inside had to remain under pressure, so the scene at the dock turned into a media event, with the aquanauts mugging through the windows and reporters covering it like a space launch gone bad. One reporter asked Simone Cousteau if she was concerned about her son in the sphere.

"I have six sons in there," *La Bergère* snapped at him. "And I am thinking about all of them."

Inside, the sphere was divided in half horizontally. In the lower half were six berths, a toilet, and storage racks for scuba gear around a moonpool. Upstairs, the six men had a comfortable galley, kitchen, food pantry, and a communications station. They were linked to the surface by telephone, a freshwater tube, and an air hose supplying a mixture of 98 percent helium and 2 percent oxygen. No human beings had ever before breathed the rarefied, unnatural mixture of gas, but at 11 atmospheres any nitrogen at all would have killed them in minutes. As it had in the Deep Cabin of Conshelf II, the helium pitched their voices up several octaves, only this time the amount of helium was so great that conversation was all but impossible. Worse, the helium seemed to dull their senses. They couldn't taste their food or smell aromas. The sea outside, rather than appearing beautiful, looked gray and sad in the eternal twilight at 400 feet. Still, they proved that divers were quite capable of tending undersea oil drilling and pumping equipment. In one sequence, two of them changed a 400-pound valve in forty-five minutes, while oil executives in Paris watched on closed-circuit television. After two days of careful decompression and the slow restoration of their gas mixture to surface air, none of them had any problems returning to the world above.

Cousteau hadn't been able to convince Columbia Pictures that Conshelf III was worth a full-length documentary so soon after *World Without Sun*. Instead, he orchestrated a live broadcast of the aquanauts' return on seventeen television stations in Europe. A week later, Cousteau flew to Washington, D.C., with thousands of feet of film he had shot of Conshelf III from *La Souscoupe,* eerie sequences of life inside the helium-filled sphere, and images of the abyss photographed by his son, Philippe. *National Geographic* executive producer Melvin Payne loved it and sold it to CBS as a one-hour special: *The Undersea World of Jacques Cousteau,* narrated by Orson Welles and edited by a hot young Los Angeles producer named David Wolper. It was scheduled for broadcast in April 1966.

The deal with CBS could not have come at a better time. Cousteau was subsisting on petroleum and oceanographic charters, money from lectures, royalties, and his share of the Aqua-Lung sales. He was desperate to find a way to push all of that aside and do nothing but make movies.

15

THE UNDERSEA WORLD OF
DAVID WOLPER

IN THE SPRING OF 1966, David Wolper sat in his living room in Los Angeles watching Jacques Cousteau's *National Geographic* special, which Wolper himself had produced, and had an epiphany that would change Cousteau's life forever.

"Look at that," Wolper said to his wife. "On the TV set, the fish look like they're in a fishbowl. This is beautiful. I'm going to get a hold of that little Frenchie."

That night, Wolper dreamed up the concept for what he thought could be years of television programs: Jacques Cousteau explores the entire world underwater. The next morning, he asked the vice president of his production company, Bud Rifkin, to fly to Monaco and find out if Cousteau was interested.

David Wolper was thirty years old when he produced his first television documentary in 1958. Born in New York City, he had migrated west to the film school at the University of Southern California. He became the business manager for the campus humor magazine *Wampus* because he thought it would be a good way to meet girls. Wolper was the only member of the staff who owned a camera, so he chose and photographed the Co-ed of the Month. He was also in charge of publicity for the USC variety show, directed by classmate Art Buchwald. To promote the show in 1948, Wolper crashed the Academy Awards at the Shrine Auditorium with a man in a gorilla suit wearing a sign around his neck that read: "If you think this is good, see 'No Love Atoll,' " which was the name of the variety show that year.

Ten years later, Wolper ran into a friend in New York who was buying cartoons produced in the Soviet Union, dubbing them with English voices, and turning a profit by releasing them in the United States. The company that was selling the rights to the cartoons was

David Wolper, creator of The Undersea
World of Jacques Cousteau
(© BETTMANN/CORBIS)

also trying to sell an hour of black-and-white footage of the launch of
Sputnik I and Laika, the dog the Russians had just sent into orbit
aboard *Sputnik II*. The dog didn't survive the trip, but it was the first
living creature sent into space. Wolper had no particular interest in
space travel or dogs, but he recognized that what the Russians were
doing and the reaction of panicked Americans was a great story. He
borrowed money from his father, bought the rights to the Soviet
footage for $5,000, hired an unknown announcer named Mike Wal-
lace to narrate, and wrote a story about German rocket scientists
responsible for *Sputnik*.

Wolper showed his film to the advertising agency that represented
Old Spice deodorant for men, which was the way television shows
were bought for broadcast at the time. They liked it, but none of
the three networks—CBS, NBC, or ABC—was interested. Wolper
had pitched his film to them through their news departments, all of
which said they would not broadcast anything unless they had total
control of the material. Since Robert Flaherty had produced the first
full-length documentary film, *Nanook of the North,* in 1922, news

departments had shunned the documentary as a form that was inferior to real reporting. Flaherty and his successors were known to stage scenes, make up situations, and add music to enhance the drama of their films. So what? Wolper replied. A documentary is the creative interpretation of reality. It is not reality.

Wolper went back to Old Spice and asked if they would sponsor his documentary if he could get independent stations around the country to air it instead of the networks. They said yes. It would be cheaper, in the long run, than paying the premium for airtime on the networks. Wolper, who had been selling cartoons and music revues to independents during the decade since he left USC, showed his documentary to the New York and Los Angeles stations with the biggest audiences. They liked what they saw. With their recommendations, he had no trouble convincing a hundred more. *The Race for Space,* with music by Elmer Bernstein, drew rave reviews and was nominated for an Oscar as best documentary film of 1958. The *New York Times,* in a front page story, hailed Wolper not only as a brilliant producer but also as the man who created his own network to broadcast his movie on television.

"I devised my philosophy of filmmaking making *The Race for Space,* and I never changed it from that first show," Wolper told the *Times.* "I want to entertain and inform, not just inform and not just entertain. I want to do both in the same piece. I saw a film in school once, and I came home and told my father. You know I saw this film in school, it was terrific. And my father said well you probably didn't learn anything. I said no, Dad, I learned more today because it was terrific. I enjoyed it, I did learn a lot. He said how can you learn a lot just watching film? And when I did *The Race for Space* I wanted to get that entertainment."

Before flying to Monaco to see Cousteau, Bud Rifkin, a veteran producer who had just sold his own company to join forces with Wolper, called Melvin Payne at *National Geographic* to ask for his impressions of the famous explorer. Payne told him that Cousteau was irresistibly charming, and immensely valuable to the world as a popularizer of the ocean and its creatures. He was a showman, not a scientist. The National Geographic Society had financed Cousteau's expeditions for a decade and given him its Gold Medal. The only problem Rifkin and Wolper would have, Payne said, was keeping Cousteau on budget. The man had absolutely no sense of what things cost and was more cavalier about financial planning than any producer Payne had ever known. Somehow, Payne added, he always makes things come out right, but it's very hard on people around him.

In Monaco, Simone Cousteau, who famously did not care for the company of most women, either liked Rifkin's wife Tedde or acted as if she did for the sake of making a deal. The two of them swanned around the Riviera while their husbands talked business. Rifkin told Cousteau that Wolper thought he could sell a television series based on the adventures of *Calypso,* its crew, and the ocean. The formula was simple. Each episode would pose a challenging question about the sea and its inhabitants. In an hour, Cousteau and his men would answer it. For instance, are sharks the vicious killers everybody thinks they are? Or how do the creatures of the coral reef depend upon one another for survival?

Cousteau and Rifkin toured Monte Carlo and the Oceanographic Museum, batting story ideas around while carefully avoiding the big question that was paramount in their minds: how much was each show going to cost? Rifkin suggested that Cousteau come up with ideas for a dozen films, fly to New York, and try to work out a deal with a network. Rifkin set the hook when he told Cousteau that if the show was a success, thirty-five or forty million people would see dolphins on a single evening.

A week later, Wolper flew to Monaco to talk to Cousteau himself. From working with him on the *National Geographic* special, Wolper had great admiration for Cousteau, believing him to be a brave man who believed passionately in what he was doing and who genuinely loved the oceans of the earth. He knew that Cousteau poured every penny he made back into *Calypso,* his museum, and his expeditions. There wasn't a mercenary bone in his body. Wolper had no illusions about Cousteau. He knew he was a tough guy who was hard on everyone around him, including his own family, none of whom, including his wife, seemed completely comfortable in his company. Wolper, a perfectionist himself, identified with Cousteau's insistence that things be done his way and done correctly the first time. They worked well together because they were equals committed to the same objective. Wolper also got along well with Cousteau because he seemed to have an air of mystery and the unexpected about him. Cousteau was gregarious and charming, but he kept his own counsel about his thoughts and desires. He was a man, Wolper believed, who though he appeared genuine in the moment, was capable of living a secret internal life.

Over dinner while working on the *National Geographic* special, Wolper had discovered that he and Cousteau shared the philosophy that poets are closer to the truth than mathematicians or politicians.

In Monaco, Wolper enjoyed renewing his connection with Cousteau. *Calypso* was another story. He inspected the ship at the dock and pronounced it unfit for duty as a television star. "It looks like shit," he told Cousteau. *Calypso* had to sparkle, and so did the divers. Black suits underwater were simply not photogenic. In his notebook, Wolper sketched out silver wet suits and yellow diving helmets with full face masks that looked like something an astronaut would wear in space. He drew streamlined plastic housings for the air tanks on the divers' backs. Cousteau's ship and his divers had to look every bit as out of this world as the NASA astronauts, Wolper said. They were the competition for airtime. Cousteau told Wolper that the cost of sprucing up *Calypso,* its divers, and its equipment would be enormous. More than $1 million. He couldn't afford to front the money. The only way Cousteau could do it was to have a deal for at least a dozen episodes, to be filmed over three or four years.

Wolper went back to New York and pitched the three networks with his idea for a series on ocean exploration built around a charismatic explorer who had already mesmerized theater audiences and won two Oscars. NBC didn't think a series about French sailors swimming underwater would hold their viewers' attention for more than a single movie-length production. CBS liked the idea, but wouldn't commit to even one season without seeing the first episode, despite having broadcast the Conshelf III *National Geographic* show that Wolper had produced. At ABC, head of programming Tom Moore said yes but only to four hours.

In the spring of 1966, Wolper still hadn't nailed down a broadcast partner. He and Cousteau continued to negotiate a production deal on the telephone. Wolper persuaded *Encyclopedia Britannica* and DuPont chemicals to sponsor twelve episodes—with the condition that a network agree to air them—and offered Cousteau $300,000 per episode. In return, Wolper Productions would own the broadcast rights for English-speaking countries and South America; Cousteau would own the rights for shows for broadcast in Europe, Asia, Africa, Australia, and the rest of the world. Wolper, who by that time was convinced that Cousteau was one of the most pleasant but hard-nosed businessmen he had ever met, suggested they get together in New York as soon as possible. Cousteau was dazzlingly brilliant, Wolper had discovered, but what made him so adept in dealing with people was a kind of primitive instinct for finding a solution to what-ever problem was at hand.

While Simone remained aboard *Calypso* for a scientific mission to measure the optical properties of seawater, Cousteau and Philippe flew to New York. During the six weeks since Rifkin and Wolper had proposed the deal of a lifetime, Cousteau and his younger son had become increasingly aware that the magnitude of the venture called for smooth collaboration between them. Cousteau's older son, Jean-Michel, clearly did not have the stage presence of Philippe, and seemed quite happy managing logistics and equipment design ashore. Philippe, however, devoured life as a filmmaker, diver, and on-camera star in *World Without Sun* and the *National Geographic* special on Con-shelf III. He was a gifted cameraman and editor, and an expert diver. At the same time, he maintained the boyish curiosity about the undersea world and its creatures that had made his father so com-pelling. As much as Philippe flattered Cousteau by emulating him, father and son often clashed furiously in differences of opinion that startled people who witnessed them. Still, there was no doubt in Cousteau's mind that Philippe would be his second in command of television production. He would eventually be his creative heir.

In New York, Philippe was not part of the actual negotiations with Wolper, but his father consulted with him daily on their progress. The rest of the time, Philippe found out what it was like to be a handsome underwater adventurer who had been on national televi-sion and spoke English with a French accent. It was spellbinding to American women. One of them, a fashion model from California named Janice Sullivan, was as irresistible to him as he was to her. They met at a party, after which they were together almost every night. Sometimes they ate dinner with Cousteau, who liked the young

American woman but pointedly excluded her from the conversation by speaking only French.

The day before Cousteau and Philippe were to leave New York, they had lunch with Wolper at the St. Regis Hotel. ABC's Tom Moore, who was sitting a few tables from them, came over to say hello. Moore was then president of the Explorers Club, an association founded in New York in 1904 whose members, over the years, had included Roald Amundsen, Robert Peary, Ernest Shackleton, Charles Lindbergh, Chuck Yeager, John Glenn, and a few hundred other bona fide explorers, along with several thousand associates who paid to join chapters of the club scattered around the world. Cousteau was not a member, but his name had come up more than once at the meetings, during which nominations were proposed. Moore, despite his apparently firm reluctance to offer Cousteau and Wolper the deal they wanted to broadcast a dozen episodes on ABC, invited them to join him for dinner that night at the Explorers Club. They finished the evening with cognac and cigars in the second-floor library, surrounded by memorabilia from the most celebrated expeditions of the twentieth century. There was a framed flag that had circled the earth in space, a ragged page from Shackleton's diary, Lucky Lindy's flying gloves. Moore never made them an offer.

A month later, Wolper got a call from Moore, who desperately needed a speaker for the Explorers Club gala at the Waldorf Astoria, which would be held in two weeks. It's a great party, Moore said, with a meal of exotic wild game and glamorous women in evening gowns. If Wolper talked Cousteau into delivering the after-dinner speech, Moore would try to persuade ABC to air twelve one-hour episodes of their underwater adventure series. It was after midnight in Monaco, but Wolper didn't hesitate. Cousteau, who sounded like he was in the middle of his workday, listened for the minute it took Wolper to outline the offer, then said one word: *Oui.*

"If Tom Moore had not been stuck for an after-dinner speaker, my father probably wouldn't have gotten the series," Jean-Michel Cousteau remembered. "When he told us about it, he said that his life was a lot of little things that came together just right."

It took Cousteau three months to disentangle himself and *Calypso* from scientific and industrial charters, including one in which his divers were helping to lay a pipeline through which an aluminum plant would discharge red-mud waste into deep water. Better, scientists reasoned, to deposit the mud in deep water, where it settled immediately as sediment, than to allow it to ruin the near-shore shallows.

In the fall of 1966, Cousteau put *Calypso* into dry dock in Marseille for a face-lift. During the fifteen years since its metamorphosis from Maltese ferryboat to all-purpose research vessel, the ship had been reincarnated many times, depending on the task at hand. Moviemaking had been a part of every mission, but rarely its sole purpose. Now, Cousteau and the shipyard crew transformed *Calypso* into the perfect underwater motion picture support ship as well as an attractive movie set capable of sailing around the world. They stripped the interior, converted the wet lab used for storing scientific samples into a darkroom, rebuilt the crew's quarters into comfortable two-berth staterooms, and added a new cabin for the Cousteaus aft of the wheelhouse. They took out the engines, generators, steering gear, and hydraulic pumps, sent them to machine shops for overhaul, and rewired the electrical system to accept shore power of the several different voltages in foreign ports. They installed two new davits, each of which could lift a ton-and-a-half motor launch, and rebuilt the large crane on the fantail to handle the recompression chamber, *La Soucoupe,* and the two new one-man subs—Sea Fleas—that had just been built by the Office of Undersea Technology. The Sea Fleas used the same water jet propulsion technology as the two-seat diving saucer, could dive to 1,000 feet, and were perfect for multiple camera shots at depths beyond those which could be reached by scuba divers. Taking out the seismic equipment for oil exploration, with its cable storage bins, made stowage room for more camera equipment, diving gear, food, and water for extended voyages. Cousteau replaced the original bow observation chamber with a new one that had much more room, eight viewing ports, and a closed-circuit television camera to send a constant stream of images to a monitor on the bridge.

As soon as *Calypso* went back in the water at the end of December, Philippe left for Paris to marry Jan Sullivan in a small ceremony with only his brother, Jean-Michel, and his grandfather, Daniel, representing the family. As Philippe's affair with Jan had deepened during the course of her several visits to France over the past nine months, Simone had made sure he knew that she and his father objected to making their arrangement permanent. A Cousteau should not marry an American, she bluntly told him, especially an American whose sole achievement consisted of wearing clothes for photographers and fashion designers. Philippe's defiance of their wishes sent shudders through the family, *Calypso*'s crew, and the rest of the Cousteau empire. The tension deepened when Simone and Jacques Cousteau refused to attend the wedding, sending as their gift a certificate enrolling Jan in an intensive French language course. After the ceremony, Philippe flew to Los Angeles to set up an office for Cousteau's production company, Les Requins Associés (Sharks Associated), near David Wolper Productions in Hollywood. Philippe then returned to Monaco to work on the first episode of *The Undersea World of Jacques Cousteau.*

Cousteau asked his older son, Jean-Michel, and his wife, Anne-Marie, to stay in Los Angeles to maintain a presence for the family while he, Simone, and Philippe were at sea. After his success at turning the institute in Monaco into a profitable oceanographic museum and aquarium, Cousteau believed such institutions had the potential not only to introduce people to the sea but to make a profit. Jean-Michel's first assignment in California was to design an underwater exploration exhibit aboard the retired ocean liner *Queen Mary,* permanently moored at a dock in Long Beach. The promoters, who were turning the giant ship into a hotel, shopping mall, and shipping museum, were delighted to attach the Cousteau name to their enterprise, regardless of which Cousteau did the work.

At a press conference in Monaco on February 18, 1967, Cousteau told the world what he planned to do with the $4.2 million he was getting from ABC television. Aboard *Calypso* with Simone, Prince Rainier, and Princess Grace at his side, he began with a passionate statement about the deterioration of the oceans caused by overfishing and pollution, which he had witnessed during the relatively brief fifty-seven years of his own life.

"I am not optimistic that the destruction can be reversed, but I am embarking on a four-year expedition to film the oceans and their inhabitants so future generations can know them as I have known them. *Calypso* and its crew will become real residents of the sea,"

Cousteau said. "On every part of the voyage, scientists from the Oceanographic Museum of Monaco and other institutions will supervise the accuracy of their discoveries."

Television, Cousteau emphasized, was the very best medium for informing the people of the world about the condition of the ocean and inspiring them with its beauty. "On a single evening in the comfort of their homes, millions of people will witness what we witness," Cousteau said in closing. The first episode would be set on the now familiar coral reefs of the Red Sea, presenting sharks as no one had ever seen them before.

At the farewell reception after the press conference, Prince Rainier and Princess Grace presented Simone with a rare St. Hubert dog. It was as big as a St. Bernard and looked like a bloodhound; though Cousteau doubted it would take to shipboard life, the dog was aboard when *Calypso* left Monte Carlo. Simone named it Zoom in honor of their sailing off on an adventure as great as riding a rocket into space.

With *Calypso*'s crew waving from the rails, the bright white ship sailed off under a canopy of confetti and balloons released by hundreds of people lining the cliffs surrounding the narrow harbor. After a brief appearance on the bow, Cousteau left *Calypso* in command of Captain Roger Maritano and gratefully retired to his cabin, where he remained for the better part of three days. Two weeks earlier, with preparations for departure reaching a fever pitch, the car in which Cousteau was riding had swerved off the road on one of his endless commutes between Marseille and Monaco. He had badly wrenched his back but refused to put off the sailing date, telling Simone that all he needed was the soothing warm waters of the tropical ocean.

The days when life aboard *Calypso* was like a party on a yacht were over. The ninety-minute *Silent World* and *World Without Sun* had taken Cousteau and his divers two years each to shoot and edit. Producing four finished hours of television in a single year was going to be an endurance test, with contractual delivery dates carrying the added weight of enormous penalties for failure. Cousteau became a taskmaster, transformed by the fact that his reputation as a filmmaker was on the line every day of his life at sea. The equipment had to work perfectly every time, and most of it was new to them because it had just been invented by the Office of Undersea Technology. In the same way that Cousteau could charm people into doing what he wanted them to do, his intensity brought anger and disgust down on anyone who failed to perform up to his expectations. The few divers remaining from the salad days of *The Silent World* and *World Without Sun* were immune from criticism, having proven themselves for a decade at sea with Cousteau. New men tiptoed around him, knowing that they could be fired in the next port if they didn't please him. And everyone noticed that Cousteau was particularly hard on Philippe.

Cousteau sent Philippe, Falco, Laban, and a team of divers to the Red Sea ahead of *Calypso* aboard the cargo boat *Espadon*. Their mission was critical: to test the two Sea Fleas, reconditioned diving saucer, new diving suits, streamlined plastic tank fairings, full-face helmets equipped with lights and radios, Galeazzi decompression chamber, electric underwater scooters, inflatable outboard skiffs, lights, and cameras. When *Calypso* arrived, everything had to be working perfectly. Cousteau's inventory of cameras, lights, tape recorders, cables, and the rest of the gear he needed to shoot high-quality motion pictures underwater sometimes made him long for the days when he, Dumas, and Tailliez had simply waddled into the sea with their Kinamo from the beach at Sanary-sur-Mer.

To film the first episode of his television series—"Sharks"— Cousteau and his crew had eighteen cameras. For topside work, they would use two 35 mm and two 16 mm Arriflex cameras, two 16 mm Eclairs, two dozen lenses, and three Perfectone synchronized sound recorders. Underwater, they would use four 35 mm and eight 16 mm Arriflexes in pressurized housings, and several Nikon still cameras in water- and pressure-proof housings. For lighting on the surface and in the water, they had sealed 1,000-, 750-, and 250-watt quartz floodlamps that could be powered by rechargeable battery packs or the 110-volt electrical system aboard *Calypso*.

The first six months set the tone for what would become a six-year epic, during which Cousteau and the *Calypso* divers would shoot hundreds of thousands of feet of film that would be cut into thirty-six episodes of *The Undersea World*. *Calypso* was still a good-humored ship, but with twenty-six men, *La Bergère*, a 70-pound dog, and all the equipment aboard, its rhythms were transformed from those of a yacht packed with energetic friends to those of a military campaign. Under French maritime law, *Calypso*'s hired crew—Captain Maritano, two mates, a boatswain, a cook, and all the divers—could remain aboard for only six months before being replaced by a second complete crew. As nominal owners, Cousteau, Simone, and Philippe were exempt, but it meant that the informal camaraderie of the early years was replaced by a much more rigid set of routines that produced thousands of feet of film every month. The plan, therefore, was to return to Marseille for crew changes and to resupply every six months.

The crew quickly learned that opportunities to film whales, sharks, turtles, and other creatures materialized suddenly and were just as suddenly gone. Like fighter pilots sitting runway alert, a camera team of divers was always ready to splash at a moment's notice unless the weather was too bad. Divers spent countless hours dismantling, inspecting, and reassembling their cameras, lights, scooters, and the rest of their gear, sometimes grumbling like soldiers forced by their officers to clean their weapons over and over.

After testing the new equipment on the familiar reefs of the Red Sea, Cousteau set a southeast course for the Maldive Islands off the tip of India, where he knew from many reports by fishermen that they would find swarms of sharks. Nearing the islands, 1,800 miles from the nearest shipyard, the starboard propeller shaft snapped at two in the morning. In the pitch-black water, with sharks gliding ominously in and out of the beams from their helmet lights, Falco and two other divers managed to lash the shaft to the hull so it wouldn't damage the rudder or fall away completely. Cousteau had a choice to make. He could head back to Djibouti or Port Sudan for repairs, or take his chances with one engine. He thought he had enough footage of feeding frenzies and menacing blue sharks from earlier voyages and recent dives in the Red Sea, but he wanted shots of the magnificent reefs and the exotic whale, tiger, and hammerhead sharks he knew he would find off the tip of India. After a raucous consultation in the wheelhouse with Dumas, Philippe, Simone, and Falco, Cousteau decided to keep going.

At 3 knots, *Calypso* limped across the Indian Ocean to the archipelago of more than a thousand coral islands. On the way, Philippe led crews in a pair of speeding Zodiac inflatable boats to film sperm whales by getting ahead of the swimming pod, stopping the engines, and keeping the cameras rolling as the giant mammals surged past them. In the calm water of the great lagoon at the center of the northern Maldives, *Calypso* divers spent two weeks baiting sharks with dead barracudas from the safety of steel cages while cameramen captured the action. They tagged some of the sharks for Dr. Eugenie Clark, who had joined *Calypso* in Djibouti as the scientist in residence to study their migratory patterns. Known among marine biologists as the Shark Lady, Clark had just accepted a job as a professor at the University of Maryland and conducted most of her research at the Mote Marine Laboratory in Florida. She met Cousteau through the National Geographic Society and had jumped at the chance to accompany him to the Indian Ocean.

In early April, still running on one engine, *Calypso* sailed for Mombasa, Kenya, where there was a shipyard that could handle the replacement of its propeller shaft. Two weeks later, with a new shaft and repairs to the frozen reduction gear that had caused it to break, Cousteau headed for the Red Sea to rendezvous with *Espadon*. He was sure he had enough footage to assemble the first episode on sharks and the second on coral reefs, but more footage would give him some insurance. He also wanted to inspect the diving saucer garage and the rest of the site of Conshelf II to find out how it had held up after five years. Most of all, he wanted to bring *Calypso* closer to the Suez Canal. For a month he had been hearing radio reports that war between Israel and Egypt was imminent. If it began before he could get back into the Mediterranean and the Red Sea ports collapsed into wartime shortages, he would be faced with long supply lines and a voyage around the Cape of Good Hope to get home.

On June 5, as Israeli planes struck Egyptian airfields and the Middle East detonated into open warfare, Cousteau talked his way onto one of the last planes out of Djibouti. Philippe, Simone, Dumas, and Falco had agreed that it would be absurd to risk their entire venture by allowing the one man on whom everything depended to be stuck for who knew how long in a war zone. *Calypso* and *Espadon* sailed north for Suez, anchored side by side on the Egyptian side of the ship channel, and waited. They immediately went on short food and water rations, since the Egyptians, who boarded for an inspection when they arrived, prohibited resupply. Caught in the middle of a crossfire, with exploding bombs and sizzling bullets all around them, Philippe and Dumas persuaded the Egyptians to let them bring some of their exposed film, cameras, and other vital gear ashore for shipment to

France. They shuttled as much as they could to the dock just before dawn on the morning of June 9, then watched from *Calypso*'s deck as Israeli F-4 Phantom jets bombed the waterfront, completely destroying the warehouse in which they had stowed the film and equipment. The next day, the warring nations declared a cease-fire. For another week and a half, Egyptian troops refused to allow *Calypso*'s crew to leave the ship. Finally, new crew members came aboard to replace the first crew, bringing with them the devastating news that the Egyptian air force had sunk dozens of tankers in the canal to deny passage to its enemies. No one could say how long it would take to clear a channel through the wreckage.

For two months, *Calypso* and its new crew stewed in the brutal desert heat. Simone reluctantly left her ship after Cousteau convinced her that she would be of greater value in Monaco, helping him come up with a new plan. Philippe went back to France, then to Los Angeles, where he and Jan, relieved to be away from his parents, put together a team of divers to film gray whales. Even though *Calypso* would eventually return to work, a second film crew on the Pacific seemed like a good idea. Cousteau and his son agreed that Philippe should be in Los Angeles when Wolper and his editors started cutting the first episode in early October for airing just after the new year. Though the tension between them was sometimes unbearable aboard ship, there was no question that they trusted each other's judgment about the film they wanted to make on sharks. Before Philippe left Monaco, he and his father also agreed to coauthor a companion book, with the help of James Dugan, to be published when the show aired.

In early September, Cousteau flew into Djibouti with Simone and crates of equipment to replace what had been destroyed during the Six-Day War. Aboard *Calypso*, he told his crew that they had no choice but to sail back into the Indian Ocean and go to work. They had lost two months, but his intention to explore and film as much of the undersea world as he possibly could had not changed. On a map on the galley table, he showed them his plan for the next two years. *Calypso* would completely avoid the Middle East and its uncertainties, and make crew changes in foreign ports instead of in Marseille. Taking advantage of opportunities to film the unexpected, they would cruise down the east coast of Africa with stops off the Seychelles,

Madagascar, and Europa Island, around the Cape of Good Hope, across the South Atlantic to Brazil, through the Caribbean to the southern tip of Florida, then to the Panama Canal. From there they would explore the west coast of South America as far south as Matarini, Peru, where they would ship their diving gear and two Sea Fleas by railroad to film underwater in Lake Titicaca. From Los Angeles, Jean-Michel was already making the arrangements for the expedition into the Andes. Back on the Pacific, they would visit the birthplace of Darwin's theory of evolution in the Galápagos Islands, then explore the entire west coast of Central and North America all the way to the Bering Sea off Alaska. Canoe Kientzy, who had taken over from Falco as chief diver on the new crew, clapped once, then again and again, until the others joined in to beat a tattoo with their hands until they burst into cheering.

16

AN HONEST WITNESS

THE FIRST EPISODE OF *The Undersea World of Jacques Cousteau* aired in January 1968. *Calypso* was at sea off East Africa with *La Bergère* aboard as always. Cousteau had joined the ship in Madagascar a month earlier to film his scenes swimming with giant turtles off Europa Island, then went back to New York for a promotional tour. In dozens of interviews, he explained himself to reporters. They knew he was a famous underwater explorer who had made award-winning movies, but wondered how he was going to entertain television audiences for a whole hour four times a year.

"My goals are many," Cousteau said.

The expeditions of *Calypso* seek to expand knowledge that would help threatened species. From cages made from Plexiglas we will film life that is serene, savage, or beautiful. We will explore the graveyards of the sea for whatever treasures are hidden, knowledge or gold. Each time we dive, each time we enter the sea, we learn something new. It is the promise that lures us, and we have never been better equipped to see, to learn, to record. This voyage is the culmination of my life's work to explore and unravel the mysteries of the sea.

Philippe Cousteau watched the premiere with his wife, Jan, in San Diego. Though the wounds of the wedding disaster continued to fester in Simone, Philippe and his father had forged a classic Cousteau truce. They left the past behind and went to work to take advantage of the greatest opportunity of their lives. For his part, Philippe compromised by enthusiastically playing second banana to his father on camera, and supervising the editing with Wolper in Los Angeles. Cousteau reciprocated by promising Philippe that after the first season was finished, he would let his son charter his own ship on the Pacific

Jacques and Philippe Cousteau during the filming of an episode of The Undersea World, *1973* (© BETTMANN/CORBIS)

to independently film an episode on whales. Cousteau, of course, would appear in enough scenes in Philippe's whale show from earlier footage aboard *Calypso* to ensure the continuity of the series.

Philippe had spent two months in Los Angeles supervising the editing of about 100,000 feet of film down to 2,000 feet for the 56-minute episode of "Sharks." He was credited as codirector with Jack Kaufman, who had worked as Wolper's assistant on the *National Geographic* special two years earlier. The script, written by another of Wolper's protégés, Richard Shoppelry, was a loosely constructed series of transitions linking dramatic footage of sharks and shots of *Calypso*'s sincere, funny, handsome, brave divers doing the dangerous work of filming them underwater. Cousteau's narration, in English with a heavy French accent, was larded with affection not only for the creatures of the sea but for his men.

"The sharks are splendid savages," Cousteau declared in the narration. "A feeding frenzy on the carcass of a dolphin is as natural to them as me eating with knife and fork. Canoe Kientzy, knowing full well the danger to himself, carefully plants a tracking tag in precisely the right place behind the dorsal fin of a tiger shark."

Tom Moore was out on a limb with *The Undersea World of Jacques Cousteau*. The real power at ABC and the other two networks was still held by the news departments. Moore was in charge of entertainment, which had been dominated for a decade by variety hours, quiz shows, and stand-up comedians. Though Moore was fascinated by Cousteau and his adventures, he had no idea how families watching television together after dinner were going to react to the first episode. ABC promoted "Sharks" as a "beautiful, terrifying study of the most ferocious creature in the ocean," putting a news report–like spin on it, not realizing that it would be Cousteau and the *Calypso* divers who would win over the audience and bring them back for more.

"One crewman rode the back of a 60-ton whale shark," wrote *Time* magazine in its review of the show. "Cousteau's red-capped divers fearlessly ran off experiments right in the menacing midst of the sharks." *Saturday Review* called the episode "unusual adventure-entertainment . . . [that] successfully combined the derring-do of divers amid dangers underwater with some meaningful marine experiment laboratory work." *Variety* called Cousteau "a master of the entertaining message." In the *New York Times,* Bosley Crowther wrote, "The intimacy with the explorers, intelligently and humorously set up, is largely responsible for the vivid sense of participation one gets from the films. At the end, you, Captain Cousteau, his crew, and their dog are friends."

A few critics, including Shark Lady Eugenie Clark, who had been aboard *Calypso* for some of the filming in the Indian Ocean, accused Cousteau of misrepresenting himself as a scientist and sacrificing accuracy in favor of theatrical impact.

"Cousteau's films are misleading in a way because they portray him as a scientist," Clark said. "I can't think of any particular scientific contributions he's made, because he just doesn't have the time. He's trapped. He needs to keep up that big image, to make it look like he's moving forward. When you get up there, when you have all that power, sometimes you lose track of what you started out to do."

"Our films have only one ambition," Cousteau wrote in an article in the *New York Times Magazine* rebutting such criticism. "To show the truth about nature and give people the wish to know more. I do not stand as a scientist giving dry explanations. I am an honest witness."

No matter what the critics said, Cousteau never forgot that story-telling was the foundation of all of his work. The overwhelming applause for *The Silent World, World Without Sun,* and the "Sharks" episode of the new television series told him he was right. Viewers wanted to see him and his crew challenged by a dramatic threat of some kind every ten or fifteen minutes or they would change the channel. They wanted layers of engagement with characters among his crew—the twinkling, rugged Falco, Philippe's handsome power, Cousteau's weathered face breaking into a smile in a shark cage with man-eaters swarming around him. They wanted the animals to be characters, too. Zoom, the ship's dog, was too big and too much trouble, so Simone replaced him with a photogenic dachshund. The dog's close-ups against plinking music signaling curiosity were cartoonlike counterpoints to the problems Cousteau and his men were having launching a Zodiac from *Calypso*'s aft deck. Cousteau devoted plenty of screen time to the creatures of the sea in their natural states, but the payoff for his audiences was their admission to the wonderful, exotic world of a bunch of French sailors with nothing to do but play in the ocean on their famous white ship.

Further ignoring the criticism that his documentaries were not really the whole truth, Cousteau orchestrated many scenes purely for effect, some of them patently absurd. On the coast of Africa, he dressed two of his men in ridiculous costumes as hippopotamuses and filmed the men in disguise trying to get closer to real hippos to film them. When he edited the footage, he played the scene as farce, and it was perfect. During the same encounter, he tried a different tack, mounting his camera on a barge, nudging it forward into the herd of hippos with an outboard Zodiac. Cousteau got the shot he really wanted when one of the terrified animals bolted for freedom and almost sank the barge on its way to open water. Almost everyone who sat down to watch a nature documentary had seen a hippopotamus, and though they were pleased to see some more, they were absolutely enchanted by the antics of *Calypso*'s crew to capture them on film, with Cousteau's seductively French-accented narration explaining it all to them.

Often, the animals in Cousteau's films were less than willing actors. He had had a change of heart after Kiki in Monaco and publicly opposed removing animals such as dolphins from their natural environment to display them as acrobats in marine parks. "It is useless to pretend that captivity in any form is less than cruel," he told one reporter. "I detest the idea of training and conditioning animals and teaching them tricks as people do in zoos and circuses." For Cousteau, the exception to that rule was capture in the name of science. If an experiment had an outcome that added to the world's understanding of wild animals, the compromise was worth it.

After the premiere of *The Undersea World,* for instance, Cousteau flew back to Port Elizabeth, South Africa, to join *Calypso* for the voyage across the Atlantic to the Brazilian coast. He was in a hurry to leave because the mild weather of the Southern Hemisphere summer was ending and his advance parties in the Caribbean already had firm dates for shooting in a dozen locations. *Calypso*'s crew had been filming sea lions on the islands around the Cape of Good Hope when Cousteau arrived. They reported that they had not gotten the close-ups and interactive shots they needed for an interesting episode because the sea lions were too wary of the human interlopers in their territory. The divers had tried going down in shark cages with no luck, and sea lions had bitten several when they didn't use the cages. Cousteau hated to waste the footage they already had of the enormous herds of the quarter-ton beasts swimming so gracefully and solved his problem by ordering his men to capture two sea lions to take with them on the voyage across the Atlantic.

The crew caught a pair of smaller sea lions and named them Pepito and Cristobal, in honor of what were to become their new home waters off the coast of a continent three thousand miles away that had been colonized by Spain and Portugal. The voyage was hell for the crew. The sea lions lived in a cage with a child's plastic swimming pool on *Calypso*'s aft deck. They ate 30 pounds of fish every day, which had to be caught, and fouled their cage and pool with an equal amount of waste that made every trip aft on the ship a nauseating ordeal. Cousteau kept the cameras rolling, filming his men hand-feeding Pepito and Cristobal and teaching them to clap their flippers as though applauding their captors.

Sea lion training stopped aboard *Calypso* halfway across the Atlantic, when word reached the ship that Daniel Cousteau had died at his home in Paris. PAC's children, Françoise and Jean-Pierre, had tended him during the last years of his life. Daniel had been ninety-three years old and ready to depart, but the news cast a pall over the rest of the voyage. For the crew, Daddy had been the ideal, doting grandfather whose diplomacy and tact had smoothed over crisis after crisis when his son had no time for details. For Cousteau and Simone, he was a patriarch who had united a family that had been almost destroyed when his sons had taken different sides in wartime France. Later, Cousteau would admit what he felt when his father died. "I did not regret my father's death," he said. "It was natural, as will be my own. I do not fear it."

Calypso reached Natal, Brazil, took on supplies, and headed for Puerto Rico. At each anchorage divers filmed the sea lions underwater, controlling them with harnesses to get them used to the cameras. Eventually, they believed, they would tame them enough to let them swim free and return to the ship because they wanted to. Cousteau built his narration for the sea lion episode around the hypothesis and conclusion in the same way a scientist frames an experiment.

"Would the sea lions follow our divers in the depths of the sea the way that dogs follow their masters for a walk through the woods?" Cousteau intoned over images of Pepito and Cristobal looking confused but cuddly in their cage on *Calypso*. "Will we be able to mount cameras on the back of one of these creatures and direct it to enter areas too small or dangerous for my divers? Can we establish friendship with these animals?" Pepito almost eviscerated one of Cousteau's divers and was put back in his cage except when on a harness. Cristobal fled and was recaptured when he surfaced next to a fisherman's boat to beg for food. Hand-fed in captivity, he had become unable to capture food in the wild.

Cousteau waxed philosophical. "Cristobal's need for freedom was something that he shared with us." He then declared success. "Two marine mammals were our willing companions in the sea. It would be useless to continue our experiment. We had already proved what we set out to prove: that marine mammals are almost as capable of attachment to humans as land animals."

"The Unexpected Voyage of Pepito and Cristobal," the fifth show in the series, drew the highest audience rating since the first episode of *The Undersea World*. Tom Moore breathed easier. He still had a hit.

The critics who accused Cousteau of showmanship at the expense of real science stung him because he knew they would hound him as long as he was telling adventure stories to attract his audience. Most scientists, however, recognized that their own passions for the natural world had found a powerful voice at the same time that television arrived to carry them to millions of people. It was a combination that could change the world.

"The captain was the key pioneer in nature filmmaking not just because he was among the first but because he recognized that productions had to be entertaining if they were to maintain the audience's attention and loyalty," said Christopher Palmer, the producer of the Audubon Society's much more traditional television series on natural history. "Television gave him an audience beyond anyone's wildest expectations."

"One of the things that people forget about my father was that he was first and foremost a storyteller," Jean-Michel Cousteau points out. "His ambition to tell stories was driven by curiosity. When he was a kid with his first movie camera the stories were about girls and himself. Later, about the ocean and its creatures because that's what he was interested in."

Cousteau's audiences grew steadily through the next three episodes, "The Savage World of Jungle Corals," "The Turtles of Europa," and "Whales," all of them shot during the first nine months of *Calypso*'s epic voyage. In the editing suite in Hollywood, David Wolper, Philippe, and JYC choreographed succeeding episodes with increasing confidence in the formula they had stumbled upon in the *National Geographic* special that aired in 1966. Cousteau's charisma; his enviable bond to his similarly attractive son; exotic submarines and diving equipment; the members of *Calypso*'s crew, whose names also became household words; and the wonderful creatures of the sea were a magic blend unlike anything ever before seen on television.

Wolper sold his company to Metromedia after he finished producing the fourth episode in the fall of 1968, leaving Bud Rifkin and a squad of successor producers to carry on. The following year's shows were even more popular than those of the inaugural season. "The Unexpected Voyage of Pepito and Cristobal" was a hilarious hour that added an unexpected dimension of sideshow humor to the series. "Sunken Treasure," about Cousteau's exploration of a Spanish galleon in the Caribbean; "The Legend of Lake Titicaca"; and "Whales of the Desert" left audiences hungry for more.

During the second half of *Calypso*'s four-year, 150,000-mile voyage that had begun in Monaco in February 1967, Cousteau's crews shot two million feet of film, with which they produced twenty-eight more episodes of *The Undersea World*. After the first season, ABC extended its production deal with Cousteau and Metromedia for $500,000 per episode, with an escalation clause depending upon audience shares. The ratings settled into a dependable pattern of ten to twelve million viewers per episode, far outrunning any other natural history series. Cousteau's formula never changed. He posed questions, met challenges in his quest to answer them, declared success even when his experiments failed, continued to capture animals to help him charm his audiences, and never failed to be the paternal presence that made learning about the sea and its creatures not only fun but important.

On July 20, 1969, the day Neil Armstrong and Buzz Aldrin walked on the surface of the moon, *Calypso* was off Unalaska Island in the Bering Sea, farther from France than she had ever been. From there, the voyage would be homeward bound. The ship's log notes that Raymond Coll took one of *Calypso*'s Sea Fleas to 500 feet, becoming the deepest man on earth while two men were the highest up there on the moon. For Cousteau, who was not aboard *Calypso* but in Los Angeles that day, the men on the lunar surface were a footnote to the miraculous photographs of the whole earth that had been sent from space by the Apollo moon ships. "Now we can see for ourselves that the earth is a water planet," Cousteau said. "The earth is the only known planet to be washed with this vital liquid so necessary for life. The earth photograph can drive a second lesson home to us; it can finally make us recognize that the inhabitants of the earth must depend upon and support each other."

Two months after the moon landing, Cousteau held a press conference at the Oceanographic Museum of Monaco to plead with the United States Congress to control the pollution along its coasts he had seen from *Calypso*. With the success of *The Undersea World*, every word he said about the ocean was being quoted in hundreds of news reports.

"The oceans are in danger of dying," he said, speaking each word in English as though it were punctuated by a period.

People do not realize that all pollution ends up in the seas. Modern fishing techniques are scraping life from the floor of the sea. Eggs and larvae are disappearing. In the past, the sea renewed itself. It was a continuous cycle. But this cycle is being upset. Shrimps are being chased from their holes into nets by electric shocks. Lobsters are being sought in places where they formerly found shelter. Even coral is disappearing. Very strict action must be taken. Some scientists are sure that it is too late. I don't think so.

The thirty-six episodes of *The Undersea World of Jacques Cousteau* that aired between 1968 and 1977 changed the way millions of people thought about the sea, but the voyage that created them had also transformed Cousteau. He still wanted to entertain his audiences. He loved the adventure of his life as an international celebrity. But now he was certain that his message about the fragile condition of the world's oceans was far more important than a pleasant hour in front of a television set.

17

OASIS IN SPACE

COUSTEAU WAS STUNNED WHEN, in 1974, ABC didn't pick up its option to broadcast *The Undersea World* beyond the spring of 1976. He had won Emmys every season and his audience still numbered in the millions for each episode. The network told him its programming philosophy had changed. The enormous success of after-dinner evenings of *Happy Days, Starsky and Hutch, Laverne and Shirley,* and the rest of the half-hour situation comedies had made ABC number one in the battle for ratings and advertising dollars. Its executives weren't willing to preempt their hit shows to air documentaries about the ocean. The ratings for *The Undersea World,* they pointed out, had dropped steadily as the networks attracted younger people who were more interested in spending a mindless half hour with two amusing girlfriends in Milwaukee than in watching a parrot fish fight to control its territory on a coral reef.

When ABC canceled *The Undersea World* and cut off the money for future expeditions, Cousteau had plenty of footage on hand to deliver the final six episodes until the contract ended. After overhauling *Calypso* during most of 1972, he had sent his ship and film crews on a two-year voyage down the east coast of South America to the unexplored waters off the Palmer Peninsula of Antarctica. *Calypso* then headed back north into the endless archipelago off Tierra del Fuego, up the coast of Chile, Peru, and Ecuador, and through the Panama Canal for a full year's exploration of the coasts, waterways, and islands of the Caribbean Sea. Cousteau had a helipad installed on the foredeck to shoot aerials and make it easier for him to get to *Calypso* for the brief scenes that gave the audience the impression he was always in charge.

Day to day, Cousteau was in constant motion. He had homes and

Jacques Cousteau, sixty-five years old (© WILD FILM HISTORY)

offices in Monaco, Sanary-sur-Mer, Paris, and Los Angeles, from which he ran the Oceanographic Museum, U.S. Divers, his undersea technology workshop, an exhibit design firm, a lecture bureau, and his most recent creation, the Cousteau Society. In 1973 he had approached a Connecticut business consultant, Frederick Hyman, about incorporating the Cousteau Group, which would own and control all his enterprises. Hyman and his corporate connections evaluated Cousteau's business plan and saw one critical flaw: Cousteau. He insisted on being in charge, but the advisers quickly figured out that he simply did not have the skill or temperament to run a multimillion-dollar corporation with several divisions. Cousteau reluctantly agreed with them.

Instead, Hyman suggested that Cousteau centralize under the umbrella of a nonprofit organization, which would limit his exposure to commercial risk while at the same time opening the way for accepting tax-free revenue and grants. They filed the corporate charter in Bridgeport, Connecticut, describing an organization whose mission was "the protection and improvement of life" and "the assumption of the role of a global representative of future generations." Cousteau was its chairman, Fred Hyman was its president, and Jean-Michel and Philippe were vice presidents. Cousteau organized an advisory board of scientists, friends, and fellow celebrities, including Papa Flash Harold Edgerton of MIT, biologist Andrew Benson of the Scripps Institution of Oceanography, science fiction writer Ray Bradbury, political activist Dick Gregory, singer John Denver, and dozens of his other famous and influential fans.

When Jean-Michel learned that his father had named Fred Hyman president of the new Cousteau Society, he resigned. Hyman, Jean-Michel told his father, seemed like a good enough man, but he was an outsider who made sense only to the family as the business manager of a profit-making corporation. For an organization like the Cousteau Society, he was the wrong choice. He had no scientific background and no status in the environmental movement, and was part of the corporation only because a head hunter had found him when Cousteau was looking for a business manager. Cousteau was furious. He prized loyalty above all else, but as always, he let Jean-Michel know that he loved him and would always have something for him to do if he came back.

By the end of the Cousteau Society's first year in 1974, 120,000 people had contributed an average of $20 each to become members. Many gave much more. Hundreds of people rewrote their wills to include bequests to the society or Cousteau himself. The popularity of *The Undersea World* was at its height, and every television hour was an unbeatable call for support. Cousteau made news constantly as he testified before Congress on a variety of hot topics, including energy, clear water, and clean air. He set up society offices in New York and Los Angeles. In Paris, he incorporated separately under French law as L'Équipe Cousteau. The staff of the society swelled to more than two hundred publicists, writers, policy analysts, artists, and clerical workers, who produced magazines, books, and brochures for a never-ending membership drive.

The Cousteau Society was a dramatic alteration of the power balance in the environmental movement that had taken hold in America and Europe in the early seventies. The Sierra Club, Greenpeace, and dozens of other international, national, and local groups had created a landscape of advocacy in which they succeeded in fund-raising with single-issue campaigns such as the slaughter of baby seals for their fur, overfishing, nuclear energy, destructive mining practices, and ocean pollution. The Cousteau Society weighed in on most of those issues but concentrated on its overarching mission of convincing the people of the world that they are dependent upon one another for survival. In the society's first press release, Cousteau wrote: "We are communicators, using words and pictures to educate living and future generations about our biological home. We are advisers, representing a kind of international State Department for the quality of life, trying to educate the world's most powerful decision makers about the ecological ramifications of their decisions."

In Cousteau's mind, all of his exploration and television enterprises served that mission, so he brought them under the umbrella of the nonprofit society. Expenses were offset by tax-free revenue from fundraising, television production, the undersea technology workshop, book royalties, and lecture fees. Cousteau wasn't worried at all about his personal finances. His salary from the Oceanographic Museum and Aqua-Lung royalties provided him with more than enough, since an open expense account covered all his expenses for the society.

Without the backing of a major television network, Cousteau and Philippe began developing a new series for the Public Broadcasting System. PBS had been created by Congress in 1967 to counteract the overwhelming influence of the three commercial networks that were broadcasting over publicly owned airwaves. Funded by federal and local taxes, PBS was a loose network of independent stations, first producing programs and broadcasting in Boston, New York, Washington, D.C., Chicago, Los Angeles, and San Francisco, and a few years later in dozens of cities around the country. With producer Andrew Solt, another of David Wolper's protégés, Philippe went to KCET-TV, the PBS station in Los Angeles, with a proposal for six half-hour programs, each of which would illuminate a single threat to the overall health of the planet. By attracting a television audience with the name of Jacques Cousteau, *Oasis in Space* would force millions of Americans to look at the population explosion, the pollution of the oceans, hunger, the devastation of chemical waste, and the dangers of nuclear power plants. The series would connect the dots for viewers with roundtable discussions among nationally known experts, interspersed with documentary footage.

Once KCET agreed to host the series, Cousteau went looking

for sponsors to pay for it. The budget for each of the six half hours shot on location around the world was about $70,000, for a total of $420,000. The national foundation that funded original PBS programming agreed to come up with $125,000. Cousteau then worked his way through a list of corporate sponsors, many of whom had happily paid for episodes of *The Undersea World*. At dozens of meetings with executives around America, Cousteau asked for money to make televisions programs that would expose truths about environmental disasters. We still love you and *Calypso,* they all told him after he outlined his six programs, but what you are talking about would be very bad for business. They make all corporations look like insensitive exploiters of a fragile planet. What's more, they agreed, not many people are going to want to watch such a dismal television show.

Cousteau ended up drawing $295,000 from the society treasury to meet budget, then watched as the corporations that had abandoned him were proved right. The first episode, "What Price Progress?," opened with shots of a Canadian pulp mill dumping mercury-contaminated effluent into nearby salmon streams and images of horribly deformed victims of mercury poisoning in Japan. Despite its grim departure from the cheery deck of *Calypso,* the first episode of *Oasis in Space* won Cousteau another Emmy. Philippe was named for producer of the best television documentary.

Ratings for the series from then on, however, plummeted. Cousteau was more convinced than ever that delivering the message that humanity had to stop fouling its own nest was his mission, but he realized that he would have to deliver that message more subtly to reach mass audiences. PBS was accustomed to the lowest ratings among the four national networks, but said it would have to see a lot more of *Calypso* and her divers or its relationship with the Cousteaus was over.

Cousteau was running out of television networks and sponsors. By the spring of 1977, membership in the society had swollen to 250,000 people, but expenses had skyrocketed. Cousteau had just closed a deal to buy a retired U.S. Navy flying boat, and his helicopter was much more expensive to fly and maintain than he had thought it would be. *Calypso* was out of commission after her vintage World War II engines had finally turned their last revolutions. He was traveling constantly, trying to raise money for a new television series he was

calling *The Jacques Cousteau Odyssey*. He promised potential investors
that each episode would be set aboard *Calypso* in the Mediterranean
Sea, combining his proven formula for underwater adventure with his
passion for protecting the health of the sea of his youth.

Cousteau was regarded as a national treasure in France, the recipi-
ent of every decoration his nation could give him, but he was most
popular among the people of the United States. His television shows
had been broadcast to audiences of millions in France, Germany,
England, Japan, and a dozen other countries around the world, but
only in America was the environmental movement shaking governments. Only in America did people seem willing to contribute millions of dollars to make that happen.

Using his network of society members, diving equipment stores,
and fans around the United States, Cousteau organized environmental rallies called Involvement Days in sports arenas in six cities—Houston; Boston; Milwaukee; Anaheim, California; Lakeland, Florida; and
Seattle. Every date was a sellout. Cousteau, Philippe, environmental
luminaries including population theorist Paul R. Ehrlich and environmental theorist Amory Lovins, celebrities including actor Jack Lem-
mon and singers Don McLean and Malvina Reynolds, and a changing
cast of national, state, and local politicians were irresistible during an
era of gas lines and foul urban air. In each city, thousands of people
joined the Cousteau Society, renewed their memberships, or simply
wrote checks. Newspaper and television news coverage kept the campaign moving. At private receptions, Cousteau appealed to philan-
thropists for donations to continue the work of the society, which was
doing nothing less than trying to save the world.

Cousteau was overwhelmed by the passionate hunger people had
for forging a new relationship with the earth and its creatures. In
Seattle, during the last Involvement Day, he and a crowd of fifteen
thousand watched a group of elementary school students perform a
primitive, powerful dance about the killing of whales. On the floor of
the basketball arena, he watched the shadowy figures of about twenty
children dressed in black move into position around what looked like
a block of fabric. Two others stood outside the circle with a white
sheet on a pole between them. When the kids settled, the rustling
silence of the arena was broken by the distinctive *racketa-racketa* of a
movie projector. The blurry black-and-white image of a ship's bow

appeared on the sheet, then a pod of spouting whales just ahead of the ship, then a harpoon gun mounted on the bow. Horrifically, the harpoon lance flew through the air trailing a rope and buried its shaft in one of the whales. After a minute, the projector ground to a halt, the two children holding the screen left the floor, and spotlights illuminated the others kneeling in a circle around a parachute that had been tie-dyed in shades of blue and green.

In turn, the children stood up and shouted the names of species of cetaceans—blue, sei, orca, humpback, gray, bottlenose dolphin, and others—their voices cracking with earnestness and stage fright. On their knees in the circle again, they shook the parachute, which rippled and gave the impression of water as soaring orchestral music blossomed from a bank of rock-and-roll concert speakers. For five minutes, the dancers rose and swam over the fabric on the floor, moving their arms in distinctly whalelike rhythms, and lifting their chins as though to breathe and spout. Their cavorting was idyllic, but then the music changed into a grating, urgent, car-chase allegro. The children began to move randomly as though confused. From the shadows on the sideline, a dancer rushed forward to the center of the floor, arms outstretched, hands together to form a point, trailing a length of rope. She struck another dancer with her hands, passed off the end of the rope, and withdrew. As the music intensified, other whale-children swam to comfort the harpooned animal. A minute later, the victim collapsed to the floor as the music fell into a mournful adagio. The other dancers gathered around their fallen comrade, lifted her over their heads on outstretched arms, settled her back on the floor, and wrapped her in the billowing fabric.

The music stopped, the children froze. The cavern of the arena was dead silent until Cousteau and Paul Ehrlich rose clapping from their seats and the audience erupted in applause that went on for ten minutes. The dancers bowed, curtsied, and smiled. Finally, the children left the floor, gathering in a locker room that smelled of sweat and wintergreen liniment, accepting hugs and kisses from their parents, who swarmed backstage. Unannounced, Jacques Cousteau materialized among them. The friends and parents moved away from the children, who fell into a ragged line. Cousteau bent to them in turn to buss their cheeks, asking which whale each child represented in the dance. They recited: orca, humpback, sperm, bowhead, blue, gray, fin . . . By the time he reached the end of the line, the children were grinning and crying at the same time. Cousteau tossed his head and laughed out loud, making a show of wiping the tears from his own eyes. He looked like one of the little dancers in black costume himself, at the center of the cluster of children, their arms reaching up to him like the tentacles of sea anemones.

As the locker room fell silent, Cousteau rose to his full height and said, "Children, you have the most important job in the world. Growing up."

The sponsor Cousteau found for the *Odyssey* series surprised him. Robert Anderson, the chairman of Atlantic Richfield Petroleum Company (ARCO), was a rarity among oil barons, a casual, seemingly absentminded man who no one would have guessed was the last of the great wildcatters. He was the son of a Chicago banker, attended the University of Chicago, where he read in its Great Books curriculum, and graduated thinking he would become a philosophy professor. His father specialized in making loans to petroleum companies, and after a summer working in the bank, Anderson was much more interested in oilmen with their tales of exploration and bonanzas in exotic climes than in the academics he met on campus. A year later, he was running a gasoline refinery in Mexico, beginning his climb to the top of an industry that had no upper limits. When Anderson's path crossed Cousteau's, he was working at ARCO's headquarters in Los Angeles, leading the assault on the North Slope oil reserves in Alaska. He was the largest single landowner in the United States. Cousteau was surprised to learn that Anderson was also attending rarefied conferences on technology, the environment, governance and social change, and Western thought. Anderson had known about Cousteau since *World Without Sun,* when every major player in the oil business was trying to figure out how to drill wells at the bottom of the sea. When Cousteau broadcast to the world the blurry black-and-white television pictures of two of his Conshelf divers replacing a valve on a wellhead at 400 feet, Anderson was one of the men watching them.

For Cousteau, there was no conflict or irony in approaching an oil business billionaire for money to produce a television series. He believed it was the duty of those who profited from the extraction of the world's resources to make sure that what they were doing to the earth did not destroy its ability to sustain life. Anderson agreed with Cousteau. Anderson also knew that in a world of soaring gasoline prices, long lines at filling stations, and year after year of record oil company profits, ARCO could use the good publicity. Attaching its name to that of an explorer and environmental crusader whose name appeared on lists of the most respected people in the world was good for business. He told Cousteau that he didn't want to see another *Oasis in Space,* and Cousteau assured him that the new series would be much more like *The Undersea World.* Anderson wrote a check for $6 million to KCET for twelve one-hour Cousteau specials. The contract named Cousteau and Philippe as co–executive producers, and specifically excluded Robert Anderson and ARCO from any decisions about content.

Anderson's check shored up the foundations of the Cousteau Society and its television production company. Cousteau turned to making sure that his always tenuous relationship with Philippe was intact. Jean-Michel continued to work on developing ocean exploration exhibits and was in demand as a speaker, but Cousteau desperately needed Philippe. With the money from ARCO and the KCET contract, Cousteau could send two separate teams to sea to produce the four finished hours of film he needed for each of the next three years. Philippe would lead one, Cousteau the other. Cousteau would fly in to both ships for his close-ups, but he was taking a giant step in the direction of an inevitable future in which his son would take his place.

At about the same time Cousteau was managing his negotiations with Robert Anderson and the whirlwind of the six Involvement Days, he was also starting a second family. Cousteau loved women. At sixty-seven he was handsome, funny, energetic, and tireless. Most irresistibly, he never concealed his passion for romance, whether for a long flirtation or a searing single night in a foreign port. He was a legend as a womanizer among his crew, who themselves enjoyed the sexual advantages of being handsome, adventurous men with charming accents. But Cousteau was famous.

During his Involvement Day swing through Houston, Texas, the local paper ran a story about Cousteau scuba diving with an unnamed woman. She had gotten into trouble in the water and had to be rescued by other divers because Cousteau was busy dealing with a bad earache. The news reports glossed over the identity of the woman diving with Cousteau, granting discretion to the male celebrity that was typical of the times.

"We are not absolutely sure," said Jean-Michel Cousteau later, "but the woman who got in trouble diving in Houston was almost certainly a woman years younger than my father who became his mistress for the rest of his life. Their life together had already begun at that point, and we know that their children together were born soon after. He kept it secret from everyone."

If JYC's natural habitat during the forties, fifties, and sixties had been the undersea world, in the seventies it was the first-class cabin of jet airliners, where he most likely met a beautiful flight attendant named Francine Triplet. Born in the landlocked Limousin region of central France, she had gravitated toward working for an international airline because of her talent for foreign languages, and was enjoying a successful career in the air. After Houston, Francine Triplet disappeared completely from public view for the next fourteen years.

18

ODYSSEY

ROBERT ANDERSON WAS TRUE to his word about staying out of the creation of the *Jacques Cousteau Odyssey* series, but the first two episodes pushed his patience. "Cradle or Coffin?" was a tedious report of *Calypso*'s five-month voyage to measure pollution and the impact of industrialization around the Mediterranean, cosponsored by the Cousteau Society, the Oceanographic Museum, and the International Commission for the Scientific Exploration of the Mediterranean Sea. Cousteau was aboard his ship for only a few hours during the voyage, spending most of his time negotiating with governments for the rights to enter their territorial waters. His crew, with *La Bergère* in de facto command, took water samples from 126 anchorages off the coasts of Monaco, France, Spain, Gibraltar, Algeria, Tunisia, Italy, Yugoslavia, Greece, Romania, Turkey, Cyprus, and Egypt. Laboratory testing of the samples for radioactive sediments, PCBs, and other toxins showed that the Mediterranean was not in grave danger but clearly a different sea from the one Cousteau knew when he was young. In Los Angeles, Cousteau and Philippe juxtaposed repetitive scenes of *Calypso*'s crew lowering drogues and sampling tubes into the sea; underwater shots of dingy, barren reefs and seafloor; and footage he had shot thirty years before in which the sea teems with life. "Cradle or Coffin?" was not the undersea adventure promised to Anderson and KCET. With Cousteau's melancholic narration, it was the most graphic depiction of the deterioration of the ocean that ordinary people had ever seen on television.

In the next episode, "Time Bomb at Fifty Fathoms," Cousteau put a much finer point on the human threat to the oceans of the world. On July 14, 1974, the 330-foot Yugoslavian freighter *Cavtat* had collided with the much larger Panamanian bulk carrier *Lady Rita* in the

Philippe Cousteau
(© BETTMANN/CORBIS)

Strait of Otranto, between the southeastern tip of Italy and the Albanian coast. *Lady Rita* survived, but *Cavtat* and its cargo of nine hundred 55-gallon drums of tetramethyl lead went to the bottom, 300 feet below. Tetramethyl lead is a flammable, colorless liquid with a slightly sweet odor that is used primarily as an additive in gasoline to prevent engine knocking. If TML, as it is known, is inhaled, swallowed, or absorbed through the skin, it attacks the central nervous and cardiovascular systems, causing nausea, convulsions, and death.

The sinking *Cavtat* was moving at 20 miles an hour when it slammed into the seafloor, rupturing its hull in uncountable places and scattering broken drums of TML into a poisonous corona around the ship. Instantly, fish, crustaceans, and mollusks within a half mile of the wreck began to die; many rose to the surface in the enormous oil slick marking the site of the collision. Two years later, after constant pressure from coastal villages whose people were afraid to eat from the sea, and an article about it in *Saturday Review* by Jacques Cousteau, the Italian parliament agreed to pay $12 million for a salvage attempt.

"Time Bomb at Fifty Fathoms" begins with Cousteau, in a plaid lumberman's jacket and his red knit diver's cap, inspecting primitive diving chambers that had been used for deep salvage attempts before the Aqua-Lung. The scene shifts to the village of Otranto, where Cousteau, now more stylishly dressed in khakis and a blue chambray shirt, walks along a cobbled street among playing children. "Around us children wheel like birds," Cousteau says, with a note of tenderness in his familiar voice. Cousteau stops in front of a half-dozen boys and asks if they remember the *Cavtat*. "Oh, yes," they say. "It is the ship that sank with the poison."

From there, a narrator with an accent similar but not identical to Cousteau's takes over with an account of the collision of *Cavtat* and *Lady Rita,* over images of the crowded shipping channel between Italy and Albania. A third narrator and the writer of most of the series, Ted Strauss, then recites in plainspoken, sonorous English the history of the battle led by a local judge that resulted in the salvage attempt. The seven or eight hundred drums of TML that did not rupture when *Cavtat* hit the bottom are ticking bombs, he says. Soon, they will rust and break open, leaking deadly poison into the sea for a generation. The only hope is to dive to the wreck, pick up the drums of TML, and bring them to the surface.

The Italian government hired an international offshore exploration company for the job. In a month its salvage ship, heavy-lift crane, and supply boat were anchored over the wreck. Cousteau arrived aboard *Calypso.* In a scene shot during lunch around the familiar table in *Calypso's* galley, the Italian judge credits Cousteau and his magazine article with turning the tide in his battle to save the Strait of Otranto, its people, fishing fleets, and tourist trade. Cousteau, dressed exactly as he had been in the shots of the children in the village, listens and nods, radiating pride and resolve. The narrator stresses that it is the judge and the people of the southeastern Italian coast who are really responsible for saving themselves and their piece of the ocean.

Over images of the wreck shot by Falco in *La Souscoupe,* with the Italian judge as a passenger, the narrator explains that death waits in orderly rows in the crumpled cargo hold. Apparently, he says, the toxins released during the initial impact have completely dissipated. The wreck is swarming with fish, barnacles, and other sea life. The scene cuts every thirty seconds or so to *Calypso's* wheelhouse, where Cousteau barks advice into a handheld microphone to Falco about the strong currents on the bottom.

For the rest of the episode, Cousteau is an honored guest, arriving

by *Calypso*'s helicopter for guided tours of the salvage ship and explanations of the difficult, dangerous work of diving to retrieve drums of
poison lying 300 feet beneath the surface. Using techniques that
Cousteau pioneered during the Conshelf experiments fifteen years
earlier, divers live in a compression chamber for three weeks, allowing
them to work at 300 feet without fear of nitrogen narcosis. They are
transported from the chamber on the deck of the ship to the wreck of
the *Cavtat* inside a steel diving bell, in which they return at the end of
their workday. Most of the time they read, listen to music on headphones, play checkers, and eat meals delivered through an air lock.
Like Cousteau's Conshelf divers, the *Cavtat* divers are tested daily for
signs of sickness or gas contamination.

On the bottom, captured on film by Falco from *La Soucoupe,* the
divers, wearing loose white biohazard suits over their dry suits, glide
like finned ghosts among the drums of poison amid the tangle of
wrinkled steel in *Cavtat*'s cargo hold. They breathe through air hoses
from compressors on the surface, and communicate with telephone
sets built into their full-face masks. The tension of men surrounded
by torn metal is magnified by a music track a lot like that from the
recent hit movie *Jaws.*

The diver in the frame wipes a glove over one of the white drums,
clearly marked with a skull and crossbones on the label. Rust stirs
from the surface. But it is not rust. The drum is leaking. The diver
swims quickly upward through his own bubbles to the open hatch of
the diving bell, sheds the biohazard suit, and climbs into the bell. In
the next scenes, two men on the surface are shown burning a load of
used biohazard suits, while others on the ship carefully wash the
divers' dry suits and masks. On closed-circuit television, a doctor
examines the diver who brushed off the contaminated surface of the
drum, and analyzes his blood passed through the air lock. He is suffering from a mild case of TML poisoning from his brief contact
with the ruptured drum. After a day off in bed, he is cleared to go
back to work. One by one, he and the other divers lift the drums into
a steel basket that can hold a dozen of them. The basket is hoisted to
the ship on the surface, where a gang of men, also wearing biohazard
suits and masks, loads them aboard. From there the TML is taken to
shore and burned.

After the last load of drums came aboard in October 1977, *Calypso*'s crew shot the final scene of the celebration as soon as Cousteau returned from the last of the Involvement Days in Seattle. All but 3 percent of the poison had been taken from the sea. The episode was Cousteau's greatest critical success in a decade, a powerful, entertaining statement of what he wanted the Cousteau Society to tell the world: the oceans are in trouble, but we humans are not helpless to save them.

Though the rest of the *Odyssey* series was nowhere near as tense or dramatic as "Time Bomb at Fifty Fathoms," PBS ratings shot up when an episode aired. In "Diving for Roman Plunder," the narration of Greek superstar Melina Mercouri saved Cousteau's hyperbolic return to the proven plot of salvaging ancient artifacts, which he had used many times before. "Lost Relics of the Sea" was a hodgepodge of shipwrecks and their links to ancient sea battles, commerce, and storms, but it was mostly *Calypso* divers doing what millions of Americans expected them to do. The two-part "*Calypso*'s Search for Atlantis" was a breezy quest for the lost continent of Atlantis in the Aegean Sea north of Crete. In it, Cousteau hypothesized but didn't really prove that the Greek island of Santorini was the former home of the lost civilization destroyed by a volcanic eruption thousands of years ago. But his failure didn't really matter. A television show about Atlantis was perfect for the late 1970s, when songs about the lost continent were on the *Billboard* charts. "*Calypso*'s Search for *Britannic*" drew a huge audience. *Britannic,* which was almost identical to *Titanic,* was sunk by a German mine off Greece four years after the iceberg disaster claimed her famous sister. Until Cousteau found the wreck, its location had been veiled in wartime secrecy, because the Germans alleged that *Britannic,* sailing as a hospital ship, had really been carrying troops and munitions. The British wanted nothing to do with proving it one way or the other, and suspiciously mismarked the location on admiralty charts. Cousteau's divers, breathing a mixture of helium and oxygen to descend to the wreck at 400 feet, found a huge gash in the hull. Cousteau and his on-camera experts solved the mystery by theorizing that the hole came not from exploding munitions but from the ignition of coal dust in the ship's bunkers.

For the two episodes they shot on the Nile River, Cousteau and Philippe left *Calypso* to showcase their latest acquisition, a surplus U.S. Navy PBY amphibious airplane. For Cousteau, adding the twin-engine seaplane to his equipment inventory gave him much more mobility and shortened production time because he could keep two or three camera crews on land and at sea all the time. *Flying Calypso,* tail number N101CS, also allowed him to share his own passion for flying with his son, who remained uncomfortable and irascible in the shadow of his father. The plane was also a peace offering to Philippe, whose love for flying far exceeded his attachments to diving and filmmaking. Philippe was already licensed to fly hot air balloons, helicopters, and single-engine light planes, and he easily made the transition to a multi-engine seaplane rating. A few years earlier, after a blowup with his father, he had come very close to leaving the family business altogether and becoming a commercial airline pilot.

In the acronym "PBY," "PB" stands for Patrol Bomber and "Y" is the code for Consolidated Aircraft, the company that built the planes. Nicknamed the Catalina by the navy, the PBY could be armed with bombs, torpedoes, depth charges, and machine guns. From its first flight in 1935, it was in military service in twenty countries for fifty years. *Flying Calypso* was a PBY-6A, the final model in a production run of about 4,500 planes built between 1936 and 1945. Unlike earlier PBYs, the 6A was fully amphibious with retractable tricycle landing gear, so it could land on a runway or on the water. It had stabilizing floats on the wing tips that folded up in flight. The two main wheels tucked into the sides of the middle of the hull, the nose wheel into a compartment that was sealed for water landings.

Powered by two 1,200-horsepower Pratt and Whitney Wasp engines mounted on a high wing, the PBY-6A had a range of 2,500 miles while cruising at a sedate 125 miles per hour. It had a top speed of about 190 miles per hour. It was not pressurized, but equipped with an oxygen system that gave it a 25,000-foot ceiling. Designed for a crew of eight, the later models of the PBY could carry twice that many people or up to 15,000 pounds of cargo. After World War II, retired military PBYs were snapped up by airlines serving coastal towns and villages without runways, medical emergency aircraft companies, firefighting teams, and private owners. For as little as $50,000, they got a yacht that slept four and could also fly.

The stars of the two episodes on the Nile River were *Flying Calypso* and its dashing pilot, Philippe Cousteau, against background shots from the cockpit and the pair of enormous glass observation blisters on the sides of the plane. The hypothesis, delivered again in the basso profundo voice of the show's writer Ted Strauss, was that after thousands of years of natural existence, human beings were affecting the course, condition, and future of the great river.

"Not only are ancient cultures and animal sanctuaries being threatened by extinction, but men are learning that technological triumphs," Strauss intoned, "sometimes create problems greater than the ones they seek to solve."

From there, the film was a natural history travelogue, with shots of the beautiful gold-and-white *Flying Calypso* over the Great Pyramids, cruising low over the sinuous, muddy river, dodging whirlwinds of insects over Lake Victoria and stampeding herds of animals below it. They rendezvoused with a convoy of Cousteau Land Rovers and headed inland for a catalog of African animals on the savanna, the same giraffes, hyenas, lions, gazelles, and the rest that everyone watching had already seen dozens of times. Somehow with a Cousteau it was better.

Backed by a symphony orchestra playing George Delarue's romantic score, Strauss's narration gave the PBY the same almost human personality that had worked so well for *Calypso* for twenty-five years. The camera spent a lot of time showing viewers how things worked, with shots of both Cousteaus and other crew members peering out into the wind of the open observation blister looking as if they were having every bit as wonderful a time as they did on their famous little ship.

During the sequences shot on land, Philippe was the central character, making decisions and interviewing locals and foreign engineers working on the Aswân Dam about the changes they had witnessed in the river. To end the show, Philippe recorded the concluding voice-over narrative, written by Strauss and usually reserved for his father: "Modern engineers, impatient for quick solutions, cannot foresee the consequences of the changes they impose," Philippe said. "Perhaps we have mastered the river. We have yet to master ourselves."

On the afternoon of June 28, 1979, Philippe Cousteau brought the Catalina out of a shallow left turn onto a northerly heading to line up with a straight stretch of the Tagus River where it flows through Lisbon and into the North Atlantic. The flight was a test hop after the plane had undergone repairs on a hydraulic pump and a routine hundred-hour maintenance check at a shop run by the Portuguese navy. As he ran through his approach checklist, Philippe savored the singular act that separates aviators from ordinary people. Almost anybody can get a plane into the air, but only a well-trained pilot can land one. He reached up and forward to a set of four control levers, set the angles of the propeller blades for maximum efficiency, reduced power, and toggled another lever to lower his wing flaps into landing position. He scanned his instruments to be sure he had fuel flow, hydraulic and oil pressure, the right engine rpm, the right rate of descent. He tapped each of three red lights on the instrument panel indicating that his main and nose wheels were safely stowed in their compartments for a water landing. If he had been approaching a runway on land, he would want those lights to be green to indicate that the wheels were down and locked. Like most pilots, Philippe repeated his final checklist once more under his breath, just loud enough to transmit into the headphones of the man sitting to his right in the copilot's seat. Flaps . . . pressure . . . three reds.

In the cabin behind Philippe and the copilot, six other men crowded around the two observation blisters toward the rear of the plane to watch the landing. The noise of the engines and the wind howling through the open windows made it impossible to talk. Below them, they watched what had been toy houses in a playroom village transform themselves into the waterfront of Lisbon as Philippe leveled off for touchdown. Skiffs and sailing smacks hugged the shore. To alert them that the plane was landing, Philippe had buzzed the river at full power before circling back for his approach into the wind. Philippe held the nose of the plane just above the horizon as it bled off speed to 75 miles per hour and settled slowly to the water. He checked one last time to be sure the river was clear ahead of him, then reached up and cut the power just as the middle of the hull skimmed the surface.

The last sensation Philippe Cousteau felt was his plane shudder, signaling its return to earth. Then nothing.

Everyone else was thrown clear. The man in the copilot seat crashed through the windshield. The others went through the glass of the observation blisters. All were cut and bruised; a few had broken bones. Boatmen who saw the plane stand on its nose and flip over fished the survivors out of the water. *Flying Calypso* sank in minutes, its wings sheared from the hull as it tumbled, the heavy weight of its engines dragging the wreckage down into the slow-flowing brown water. Only Philippe was missing.

For two days, they looked for his body. By the end of the first day, the family had arrived from Los Angeles: Cousteau; Simone; Philippe's wife, Jan, with their daughter, Alexandra, and pregnant with their second child; Jean-Michel; PAC's son, Dr. Jean-Pierre Cousteau; and several friends who traveled with them for support. Dozens of journalists flooded into Lisbon to report on the search. Overcome with emotion, the Cousteaus could do nothing but wait in seclusion in a villa owned by Air Liquide while a Portuguese military helicopter searched the shoreline downstream and thirty navy divers groped in zero visibility along the bottom of the muddy river. It didn't take long to find the cockpit. Philippe was not in it. When they hoisted the wreckage onto the salvage barge, it was obvious that something catastrophic had happened to the left front side of the plane's nose. Unlike the rest of the wreck they recovered, the metal on the pilot's side was torn and wrinkled, unrecognizable as part of an airplane. The fear set in that whatever had happened in the crash had so terribly destroyed Philippe's body that his remains would never be found.

Finally, on the morning of the third day, Dr. Cousteau, who was handling communication with the world outside the villa, took a phone call from the chief of police. Divers searching under the waterfront wharves had found Philippe pinned almost invisibly against pilings by the current. There wasn't much left that looked like a man, the policeman said. He had consulted with aviators on the scene who seemed to think that the left engine had broken loose on impact and slammed into the cockpit right where Philippe was sitting. No one had any idea what caused the crash. Perhaps it was a nose wheel door that had somehow come open and pitched the plane forward when it struck the water at 80 miles per hour. Or the plane might have hit a sandbar or some other submerged object. It was impossible to know for sure. Someone, the chief concluded, had to come to the city morgue to identify the body. Jacques Cousteau choked out loud as though keeping himself from vomiting and left the room. A moment later, he called back to Jean-Michel. You must do it, Cousteau said. I cannot.

Philippe's funeral was at Lisbon's Saint-Louis-des-Français church and his burial at sea from the Portuguese corvette *Baptista de Andrada*.

With the family and a camera crew aboard the warship, Cousteau took the body of his son 25 miles out into the North Atlantic and sent it into the embrace of his beloved ocean.

Afterward, on the dock at the navy base, Cousteau spoke for the first time about his son's death to the crowd of several dozen journalists representing the major papers and news organizations of the world. "Nothing is changed in our program," he told them, shocking some of the reporters and many of *Calypso*'s crew who read their stories. Cousteau's opening remark seemed brutally crass in light of the fact that he had just buried his son. Everyone who knew Cousteau understood that he was not a man to dwell on the failures and tragedies of the past. He was a master at moving on without reflection or regret. But this was Philippe.

"Our upcoming expeditions will proceed as planned," Cousteau said. "What was a tragedy for my son was a miracle for his copilot. The propeller that killed Philippe saved the life of the copilot, who was tossed, uninjured, to the surface. That is fate. We must accept it and go on. I have another problem now. Philippe, of course, was to take over and continue my work when I am gone. There must be someone to run the society. There must be continuity."

After that day, Cousteau refused to talk publicly about Philippe ever again. Six months later, he wrote a letter to his dead son and published it in one of his reports to the members of the Cousteau Society.

> *Mon cher Philippe:*
>
> *I will always remember that day when you joined our Conshelf Two expedition. I was impatient to show you our Village under the Sea before it became too dark. Hastily, we submerged. I kept your hand in mine, to guide you. I felt strangely proud, not of what we had achieved, but because our dreams were always shared so intimately.*
>
> *Three years ago, I found myself sitting near you in the cockpit of your Catalina. I looked at you, my guide in the sky as I had been your guide in the sea. I saw your shining face, proud to have something to give back to me, and I smiled because I knew that pursuing rainbows in your plane, you would always seek after the vanishing shapes of a better world.*
>
> *I love you, JYC*

19

MOVING ON

AFTER LISBON, SIMONE RETREATED to the familiar routines of
Calypso, which were all she could bear of the world. She got very little
comfort from her husband anymore, so the pain of Philippe's death
was best endured alone at sea. Aboard her ship, Simone enjoyed the
privacy of her stateroom behind the wheelhouse when she wanted it,
and the camaraderie of the galley at mealtimes if she chose to eat and
drink with the crew. The Venezuelan government had chartered the
ship to sample the ocean at the mouth of the Orinoco River, a three-
month job without the hectic pace of film production. There were
cameras aboard, as always, and Cousteau appeared for two days to
shoot a few scenes, but the mission of the voyage was straightforward
and relaxed.

Away from *Calypso,* Cousteau threw himself into a maelstrom of
plane trips and meetings, editing the last three episodes of the *Odyssey*
series in Los Angeles. He continued negotiations, on which he and
Philippe had been working, to change *Calypso*'s home port from
Monaco to Norfolk, Virginia. There was no question in Cousteau's
mind that closer ties to the United States were in the best interest of
his family, the society, and his other enterprises. Norfolk was the
home of the American fleet and countless shipyards, and was one of
the best-protected harbors in the world. It was only 150 miles from
Washington, D.C., the heart of international environmental advocacy.
For Cousteau, perpetual motion was as much of a balm to the agony
of losing Philippe as *Calypso* was for Simone. In a single month that
fall, he was in Norfolk, Paris, Los Angeles, Norfolk, Paris again,
Monaco, Venezuela (to look in on *Calypso* and Simone), and, again,
Los Angeles.

Before Philippe died, Jean-Michel had accepted his life on the

Jean-Michel and Jacques Cousteau (PRIVATE COLLECTION)

fringes of his father's enterprises. He had carefully used the Cousteau name to build a business for himself as a popular lecturer to audiences of environmentalists, marine scientists, and divers. His father continued to turn to him for advice on creating exhibits, including the early stages of a plan to build a Disneyland-like ocean park in France. After Philippe died, there was no question that Jean-Michel would return to his father's side and do whatever needed to be done. The moment he heard about Philippe, Jean-Michel knew it was time to close ranks, to come together. His father told him he could not go on. His work was over. Jean-Michel said, "JYC, I'm in. Don't say another word. It's all taken care of."

Jean-Michel, with his wife, Anne-Marie, and their children, Fabien and Céline, moved to New York to run the Cousteau Society and *Calypso*'s expeditions. When Cousteau introduced Jean-Michel in his new role, he said, "He is an architect but also a teacher of marine biology and an ecologist. He cannot replace Philippe, but he has other talents. He is a much better administrator than either Philippe or I could ever be."

Jean-Michel punctuated his father's estimation of his abilities with his own remarks to the press as he packed up to leave Los Angeles.

"With the death of Philippe we've lost a talent I don't have," Jean-Michel told reporters. "Philippe was full of poetry and dreams, he was a beautiful storyteller, a talented filmmaker. His world was in the air. How can we reinstate that perspective? I'm not a balloonist, not a pilot. I'm an ocean person. The society will always be marked by his absence."

Editing the film for the last three episodes for PBS was very nearly more than Cousteau could bear, because his lost son was in so many frames. Jean-Michel took over much of the work. For "Lost Relics of the Sea," they cut previously shot footage into a catalog of shipwrecks in the Mediterranean and Caribbean. He edited "The Warm-blooded Sea: Mammals of the Deep" from a decade of film and a script by his new writer, Susan Schiefelbein of the *Saturday Review*. Her story began with a poignant encounter with fur seals and their pups, the icons of a campaign against the hunting of marine mammals, and traced the evolution of seals, whales, and dolphins. Cousteau and actor Robert Wagner narrated, with a classically inspired orchestral score. Shots of Philippe swimming with a tame dolphin, Jean-Michel in the water watching dolphins feed on a school of menhaden, and Japanese fishermen slaughtering dolphins to keep them out of their nets were combined, indicting the film's audiences as much as entertaining them. Time is running out for the seal pups, was the message, and it is running out for all of us, their relatives, too.

"Why can we only point accusing fingers and never hold up a mirror?" Cousteau asked in his narration.

PBS, ARCO, and Robert Anderson started feeling uneasy about continuing the *Odyssey* series for another season after those two shows. The final episode, "Clipperton: The Island That Time Forgot," left no doubts that they were finished with Jacques Cousteau. Philippe had led a six-week expedition to Clipperton in 1970 when he was trying to distance himself from his father with his own production company. He shot thousands of feet of film of accompanying scientists examining the unique coral reef ecology surrounding the isolated island, 700 miles southwest of Acapulco, but never produced a finished film. Ten years later, Cousteau and Jean-Michel used Philippe's footage as background for their own graphic investigation into the story of a lighthouse keeper on Clipperton who had brutalized a harem of women and children until they rebelled and killed him with a hammer. When Cousteau returned to the island to shoot scenes for the *Odyssey* episode, he brought back one of the surviving children, who described the murder on camera and erected a cross at the scene of the crime.

Despite Cousteau's heroic attempts to soldier on, his subconscious betrayed him in "Clipperton." He and Jean-Michel edited Philippe's underwater shots into scenes that echoed and amplified the human tragedy that had taken place on the island. Sardines fleeing upward from a shark attack below are finished off near the surface by tens of thousands of blue-footed boobies. Swarms of crabs scuttle to the tide line in a frenzy to grab the sardines dropped by the birds. Moray eels wriggle ashore to eat the crabs. Even the water itself is deadly. A freshwater lagoon is filled with decaying plants, poisoning the water with chemicals that eat through the divers' suits and masks, burning their eyes and skin. In the narration, Cousteau says, "My skin was attacked as if immersed in acid. It became intolerable. When we surfaced from the inferno and took our masks off, the bad smell we had carried with us was suffocating. Our yellow tanks were bleached by hydrogen sulfide, the metal parts of our Aqua-Lungs were black as coal, and our red eyes leaked tears for the rest of the day."

The *Los Angeles Times* television critic pointed out that the underwater scenes were, if anything, more melodramatic than those of the survivor of a night of murder arriving back on a remote island and kissing the ground as he stepped ashore. Other critics were far less kind. The *New York Times* ignored all three of the last episodes of *The Jacques Cousteau Odyssey.* The *Washington Post* pointed out that Cousteau was among the most celebrated explorers and most famous entertainers of all time, and had suffered an unimaginable tragedy with the loss of his son. He was entitled to a rest. PBS programming executives agreed with the *Post.*

A month after Cousteau finished the last episode for PBS, he moved his editing studio from Los Angeles to Paris as part of another major change in the way he produced television shows. Until then, he had steadfastly clung to shooting film because of the higher resolution, better color, and his familiarity with it after sixty years of using movie cameras. The technology of video recording had improved so much, however, that he could no longer justify the intermediary step of physically assembling film before transfer for electronic presentation on television. Cousteau told anyone who asked him that he was moving his production studio to Paris so he could use the European Phase Alternating Line (PAL) video system, which operated on 50-cycle current. It produced crisper images than the American National Television System Committee (NTSC) format used in the United States, which operated on 60-cycle current. Leaving Los Angeles also meant leaving memories of Philippe, who had firmly set his anchor on the California coast.

With his studio in Paris, Cousteau was, not coincidentally, in the same city as Francine Triplet. Though their life together was a secret, they had two children by this time. Simone was always aboard *Calypso,* saddened beyond redemption by Philippe's death, so Cousteau's second family had become his home. His parents were dead. His son Philippe was dead, his wife estranged though cordial. Only Jean-Michel was part of his daily life.

Jean-Michel's arrival at the Cousteau Society was not cause for celebration among the staff. He went through the books and discovered that he had inherited responsibility for a swollen, inefficient bureaucracy with dramatically declining revenue after the cancellation of the PBS contract. Even with its membership at almost three hundred thousand, the Cousteau Society had somehow fallen more than $5 million in debt. His father and Philippe had shared a similar philosophy about money: spend what you have; get more when you run out. To a methodical thinker like Jean-Michel, the society was facing a financial crisis that could force it into bankruptcy. Within months, he had trimmed 20 percent of the staff and shaved hundreds of thousands of dollars off the operating budgets of the several dozen separate enterprises under the wing of the society. While he attempted to bring order to the ledgers, he shipped millions of feet of movie film from Los Angeles to Paris, and moved expedition equipment from Marseille to Norfolk along with the society's membership and publication offices, which had been in New York.

Jean-Michel was above all else a builder. He relished the idea of creating a grand headquarters for the Cousteau Society in Norfolk, envisioning offices, warehouses for supporting *Calypso,* and a museum celebrating his father's adventures. The politicians and business communities of Norfolk were similarly enthusiastic. Several cities on the Atlantic coast had competed for the privilege of becoming Jacques

Cousteau's home port, and Norfolk outbid the others with an offer of $75,000 to cover moving expenses, free office space, and a free dock for *Calypso*. That much of the deal was done, but the far more ambitious redevelopment of the waterfront, for which Cousteau wanted an additional $5 million to build his Ocean Center and Museum, was still not settled.

After Jean-Michel and Cousteau consolidated in Norfolk and brought the society budget under control, they still faced a cash shortage. *Calypso,* back from Venezuela, was docked in her new home port through the spring of 1980. During the city's annual harbor festival in April, thousands of visitors streamed aboard to tour the famous ship, some of them wondering why it had been at the dock all winter and not off on a fabulous new adventure.

The adventure, when it materialized a month later, was launched by a million-dollar grant from the Canadian Film Board to produce two hours on the St. Lawrence Waterway for broadcast on PBS in Canada and the United States. Cousteau arrived from Paris to lead the expedition from Norfolk harbor in early June. *Calypso*'s departure made network news along the eastern seaboard with film shot by a society camera crew of Cousteau, Jean-Michel, Albert Falco, and a new generation of *Calypso* divers waving from the rails. Under the customary canopy of green and white balloons, with all flags flying and an escort of whistling tugs and launches, *Calypso* sailed into the North Atlantic and turned north for Halifax and the Gulf of St. Lawrence. Four hours later, Cousteau left his ship by helicopter to connect with a Concorde flight back to France.

Even with his mother to smooth the way aboard *Calypso,* Jean-Michel had trouble assuming command. Falco was the expedition leader, Alaine Traounouil was the captain in charge of running the ship, and everybody else aboard seemed to know more about what was happening and how to do it than Jean-Michel. The Canadian film crew that joined the underwater cameramen for the voyage were stunned when they witnessed the first day of shooting one of Cousteau's fly-ins for his close-ups. The camaraderie of the crew during the scenes seemed forced for the sake of the cameras, and Cousteau's fondness for his own celebrity was unmistakable. It was as though Hollywood had come aboard, and nothing about that day in Halifax resembled anything they had seen on television.

From Halifax, *Calypso* sailed east, stopping at Sable Island, 95 miles offshore, to film shipwrecks, and off the islands of St. Pierre and Miquelon, the lone remnants of the French colonization of North America. While circling Newfoundland they came across a baby humpback whale tangled in a fishing net with its mother circling nearby. Neither Cousteau nor Jean-Michel was aboard, but old hands Bernard Delemotte and Raymond Coll led a team of divers into the water to cut the little whale free. It took them a half hour longer than it should have because neither Delemotte nor Coll wanted to report the incident to Cousteau without film in the can. The scene was touching and ripe with tension. Delemotte stroked the young whale and wondered aloud if he could cut away the net without injuring the whale. Once free, the baby whale swam on the surface for a mile with Delemotte on its back holding on to the dorsal fin. Then the mother and child followed *Calypso* until nightfall in a poignant ending to the encounter that fit perfectly into Cousteau's view of marine mammals and humans sharing the natural world.

After skirting the east and north coasts of Newfoundland, dipping into fjords and bays to explore and film bird colonies, shipwrecks, and marine life, *Calypso* entered the Gulf of St. Lawrence, the easternmost of the chain of waterways that connects the Atlantic to the Great Lakes at the center of the North American continent. Well into the winter of 1981, with rare visits from Cousteau, *Calypso* roamed the Great Lakes. In September, he arrived for a welcoming celebration in Detroit when a flotilla of four hundred boats escorted *Calypso* to the city on the American side of the border. That visit was bittersweet because just a week earlier, thirty-year-old diver Remy Galliano had died of an air embolism on a routine descent to a shipwreck in Lake Ontario. Cousteau came back to his ship again in early November after a bitter cold spell had crippled *Calypso* in an ice storm, filming scenes in which he chipped ice from the rails and rigging with the rest of the crew.

Jean-Michel spent much more time aboard *Calypso* than his father. He led an overland expedition to explore beaver dams; dives to several shipwrecks, including that of the *Edmund Fitzgerald* in Lake Superior; and a snowmobile excursion up the glaciated face of a hydroelectric dam in Hudson Bay. He left the ship in Montreal in late November, while a shortened crew, including Simone, made the monthlong trip back through the St. Lawrence Waterway, down the coast of New England, and into Norfolk to repair ice damage to *Calypso*'s hull and engines.

The film of the voyage that Cousteau and his editors in Paris cut for broadcast in the fall of 1982 was more of an inventory of *Calypso*'s equipment and technology than the natural history of the St. Lawrence expected by the Film Board. The whales, beavers, birds, and ship-wrecks were strung together without any real links at all except for endless shots of the helicopter, *La Souscoupe,* Land Rovers, Zodiacs, and a hovercraft performing for the cameras. The crew no longer seemed to be the lucky-and-we-know-it band of adventurers, but rather a mil-itary unit weary from too much work and no days off. Critics panned the two episodes of *St. Lawrence—Stairway to the Sea* when they aired. The audience ratings for the shows on the Canadian Broadcasting Company and the few PBS stations in the United States were among the lowest ever for a Cousteau production. It was as though, one critic said, the Cousteaus came to Canada to be congratulated on being the Cousteaus rather than to reveal what they found in a meaningful way. One Canadian on the camera crew said he thought Cousteau had been content to use his famous name to get the Film Board's million dollars.

While *Calypso* was on the Venezuela and St. Lawrence expeditions, Cousteau spent much of his time ashore on a venture on which he and Philippe had been working since the oil shortages of the mid-1970s. If the world was running out of oil or if oil was going to become prohibitively expensive, he reasoned, why not return to wind power to propel ships? Cousteau decided that his replacement for the aging *Calypso* should be a test platform for an improved version of a rotor sail system pioneered in the 1920s but never perfected. The Magnus rotor, invented by German engineer Anton Flettner, could theoretically propel a ship with a vertical wind-driven drum turning at a speed of 200 miles per hour. The rotation created a partial vac-uum on the front side of the drum, sucking in air and moving the ship by using the same principles of lift as those of an airplane wing. The problem was that at 200 rotations per minute, the drum was as deadly as a giant meat slicer. Flettner had given up.

Cousteau approached aeronautics professor Lucien Malavard, who had helped design the Concorde, and asked him to lead a team in preparing a request for a grant from the French government to develop a practical rotor sail. In September 1980, with the panic of expensive, scarce oil lingering like a haze over international commerce, France again went into business with Cousteau. He and Malavard received a million-dollar grant to perfect the Turbosail. Cousteau leveraged the grant into backing from Pechiney, a French metals conglomerate, with assurances that the prototype Turbosail ship would be built of aluminum, as would the Turbosails themselves. After demonstrating the concept, Cousteau told them, he could retrofit existing freighters of the class from 3,000 to 80,000 tons with Turbosails, run them in tandem with standard diesels, and save 35 to 40 percent on fuel costs. Propulsion control systems had already advanced far enough to manage both forms of energy by computer. Pechiney liked the idea but wanted to own the patents. In exchange for a portion of the royalties payable to the Cousteau Society, Cousteau agreed.

A year later, in a wind tunnel near Toulouse, Malavard demonstrated a working Turbosail for Cousteau and Pechiney executives. It was a 44-foot-high hollow aluminum column with a parabolic leading edge and a semicircular trailing edge, aerodynamically similar to an airplane wing. It generated forward force by directing air through the cylinder and producing a drop in air pressure on one side and an increase on the other. Malavard and his engineers had also figured out a way to link the Turbosail and the main engines of a ship to maintain a constant speed regardless of wind.

Cousteau bought a 65-foot catamaran, renamed it *Moulin à Vent* (*Windmill*), and had Malavard install his nonrotating Turbosail on its foredeck. In October 1983, *Moulin à Vent* was ready to sail from Tangier after testing in the Mediterranean, during which it had reached speeds of up to 10 knots under Turbosail alone. Cousteau instructed the Cousteau Society staff to prepare a gala welcoming ceremony in New York for mid-November, specifically requesting fireboats and fireworks for his entrance.

With Cousteau, Jean-Michel, and a crew of five, including a cameraman and gaffer, *Moulin à Vent* sailed for America. *Calypso*—and Simone—sailed north from Norfolk to make the entrance into New York Harbor together. For ten days, the voyage across the Atlantic was unremarkable except for the boredom and cramped conditions aboard the little ship. Then, 400 miles southeast of Bermuda, they sailed into a gale. In 50-knot winds and 20-foot seas, the Turbosail began to tear loose from the deck. Cousteau shut down the rotor, reinforced the stays and guy wires, and diverted to Bermuda for repairs. A week later, a day out of New York again in rough seas, the Turbosail broke completely free of the deck and tumbled into the sea, barely missing the cabin. Cousteau canceled the welcoming ceremony and took his crippled windship to Norfolk, where he held an impromptu press conference on the dock.

"We have lost only the hardware," he told reporters. "The brains who have conceived the systems are already at work. Give me a little time and we will do it again. This is not dream stuff. This is economic reality." His next windship, he promised, would be twice as large as *Moulin à Vent* and have two Turbosails.

20

CAPTAIN OUTRAGEOUS

AS THE ME GENERATION status seekers of the 1980s drowned out the altruism and environmental awareness of the previous decade, Jacques Cousteau somehow remained at the top of all the lists of the most recognizable faces in the world. The waning popularity of his television shows did nothing to reduce the magnitude of his fame, which he used to advocate for population control, nuclear disarmament, and the protection of rivers and oceans. Financially, Cousteau was teetering on the brink of catastrophe yet keeping the throttles on most of his enterprises wide open. The bill for repairing *Calypso* after the ice damage from the St. Lawrence expedition came to more than $100,000. His windship, *Moulin à Vent,* was a derelict on the Norfolk waterfront. At Cousteau Society headquarters at 777 East Third Avenue in Manhattan, the very real possibility of bankruptcy soured the workdays. Cousteau and Jean-Michel had no choice but to call a meeting of the society advisory board to construct a vision of the future that might not include the production of films for television. PBS and the three networks had emphatically slammed their doors. The French government was running as fast as it could from the Turbosail project. Only a relative trickle of memberships, renewals, and donations was keeping the society afloat. Despite their dire financial situation, Cousteau and Jean-Michel enthusiastically introduced their plan for a five-year expedition to the Amazon, the South Pacific, and the rivers of Asia.

After a particularly bitter session during which an accountant pegged the Cousteau Society's debt at $5.1 million, John Denver, who had been a loyal adviser since the beginning, pulled Jean-Michel away from the dismal group. Denver said he was working on his own music special on the new television network owned by Ted Turner, who had recently emerged as a celebrity in popular culture when he

John Denver, Jacques Cousteau, an unidentified woman, and Ted Turner
on Calypso, *celebrating Cousteau's seventy-fifth birthday in*
Washington, D.C. (COURTESY ROGER NICHOLS)

won the America's Cup yacht race. Turner is a little bit different, Denver said. Kind of an odd duck, a bit of a rager, but a powerful man with some very interesting and enlightened ideas about the ocean. Denver told Jean-Michel that he'd be happy to arrange an introduction. Jean-Michel led Denver to an empty office and told him to call Turner right now.

Robert Edward Turner III was the son of an advertising man who had specialized in billboards to build a business worth more than $1 million by the time he committed suicide in 1963. Ted inherited Turner Outdoor Advertising after his father's death. Twenty-five years old, he had been expelled from Brown University three years earlier for having a woman in his dorm room. Afterward, he spent most of his time racing sailboats out of a yacht club in Savannah, Georgia. Five years after Ted took over, Turner Outdoor controlled the billboard market in northern Georgia and southern South Carolina, owned a half-dozen radio stations, and was looking for a television station to buy. He heard that Channel 17 in Atlanta was losing $50,000 a month with less than 5 percent of the city's television viewers watching its programs. Turner had plenty of cash and thought it was a bargain compared with the price of a network affiliate. Federal law now required that UHF frequencies be built into new television receivers, along with the thirteen familiar channels broadcasting in VHF. What difference did it make, Turner figured, if a channel's number was 17 instead of 4, 5, or 7? Why did NBC, CBS, and ABC mean any more than the Turner Broadcasting System if he could transmit a signal into every home with a new television?

Turner bought more television stations around the South, but WTCG-17 was his masterpiece. In 1976, with some more of his cash, he gave the people of Georgia a compelling reason to tune into WTCG-17 when he bought the Atlanta Braves baseball team and the Atlanta Hawks basketball team. After that, if you wanted to watch local baseball or basketball you had to watch it on Turner's channel. The same year, he got permission from the Federal Communications Commission to bounce his signal off a satellite and down to local companies that sold Channel 17 as part of a package of channels piped into homes through cables instead of over the airwaves. Cable television offered not only more channels but clearer pictures. Ted Turner's Channel 17, renamed WTBS (W Turner Broadcasting System), became America's first cable television superstation. His programming consisted of Braves and Hawks games, reruns of other sporting events, old movies, sitcom reruns, and cartoons, all of which he either owned outright or bought for next to nothing.

In a year and a half, WTBS was worth $100 million. Turner was being celebrated in *Fortune, Time,* and the rest of the front rank of American media as the country's most successful swashbuckling entrepreneur. He became an even more prominent international celebrity in 1977 when he bought a three-year-old aluminum 12-meter yacht that was supposed to be a sparring partner for two new boats competing to defend the America's Cup. Turner's boat, *Courageous,* had won under its original owners three years earlier, but the newer *Enterprise* and *Independence* were supposed to be much faster.

With Turner himself at the helm, *Courageous* beat the other American yachts in challenge races and took the best-of-seven series against Australia's challenger *Australia* in four straight races off Newport, Rhode Island. In a sport awash in egos, Turner bested them all with braggadocio, taunting, cigar smoking, and drinking against the glitzy background of international yacht racing. The media, delighted as always by a colorful showman who made good on rash promises, dubbed him Captain Outrageous and the Mouth of the South.

After John Denver's introduction, Turner and Cousteau sat down to talk in the summer of 1981. They liked each other immediately. Jean-Michel, who went to Atlanta with his father, reported to John Denver that their meeting was love at first sight. Both men were cultural icons who radiated confidence as if they had invented it. Instead of clashing, as two such similar men usually do, their pragmatism and mutual respect prevailed. The adventures of Jacques Cousteau, recent gas shortages, and the nuclear disaster at Three Mile Island had inspired a quirky environmental awareness in Turner. He was a runaway capital-ist and developer, but contributed millions of dollars and free televi-sion time to promote population control, solutions to world hunger, and safe, nonnuclear energy sources. He believed that a single man like Cousteau who could deliver important but at the same time entertain-ing messages to the world was more essential than a thousand scientists who made sense only to one another.

Though Cousteau betrayed nothing, he was slightly uncomfortable approaching Turner. He was a seventy-one-year-old filmmaker who was $5 million in debt, and he was not at all convinced that cable television was the way forward. Turner immediately set Cousteau at ease with a gracious acknowledgment of Cousteau's contributions to humanity's relationship to the environment, his skill as a filmmaker, and his value as a guide to the beauty of the natural world. In person, Turner was soft-spoken and thoughtful, nothing like the raucous sailor he appeared to be during the America's Cup races. He quickly convinced Cousteau that in the same way that broadcast television had been a better means of reaching audiences than theatrical films, cable television, with much clearer pictures paid for by individual subscribers instead of advertisements, was the future.

With the preliminaries out of the way, Turner asked Cousteau what he wanted. Cousteau was brief and to the point. The great rivers of the world upon which the oceans and all life depend for sur-vival were becoming toxic sewers. He wanted to continue what he and Philippe had started on the Nile and make a six-month voyage on the Amazon, from which he would produce four hour-long televi-sion specials. How much money do you need? Turner asked. Six million, Cousteau said. Turner stuck out his hand to shake. You've got it, he said.

While Cousteau and Turner worked out the details of the contract, Jean-Michel negotiated with the superstation programming staff. The Amazon shows would not be ready for broadcast for at least three years, but the Cousteau Society had retained ownership rights to all twelve episodes that had aired on PBS from 1978 to 1980. The next day, after consulting with Turner, WTBS bought the exclusive rights to *The Jacques Cousteau Odyssey* series for $5 million, payable over the next five years. In New York and Norfolk, the fog of financial desperation that had settled over the Cousteau Society and the Cousteau family began to lift.

The exploration of the Amazon was Cousteau's most complex and expensive expedition. For fifteen months, while he negotiated with the governments of Brazil, Peru, Bolivia, Guyana, Surinam, and Ecuador for rights of passage, Jean-Michel managed a team of fifty men and women preparing for the voyage. Albert Falco and Raymond Coll, the last of the original *Calypso* divers, organized underwater camera teams, trained new divers, and supervised the building of their equipment inventory in Norfolk. Jean-Michel's wife, Anne-Marie, was the expedition's still photographer. Susan Schiefelbein went to work on the telescripts. Cousteau Society staffer Paula DiPerna set up an office in Manaus, Brazil, as the logistical ringmaster in charge of hotel rooms, plane tickets, boat charters, medical contingencies, and equipment shipments.

Because *Calypso* was going to be far from a shipyard for at least two years, she had to be in perfect condition when she left Norfolk. Through the spring and summer of 1981, Falco, Coll, and their crew worked with a handful of professional shipwrights to replank the ice-gouged bow, repaint and recaulk the rest of the hull, and rebuild the rudders, propellers, and shafts to original condition. Chief engineer Jean-Marie France directed the complete overhaul of both main engines and the removal of the twin auxiliaries that powered the electrical and hydraulic systems. He replaced the old auxiliary engines with a pair of brand-new model 6-71 diesels that were a gift to the society from the employees of the Detroit Diesel Allison factory that built them. On one of the new engines, the factory workers had fastened a plaque honoring the environmental contributions of Cousteau and his ship.

The mission of the expedition was straightforward. Jean-Michel would travel overland and on smaller tributaries from the Amazon headwaters, and rendezvous with *Calypso* working its way upriver from the east. Both teams would concentrate on finding out how the human presence was affecting the Amazon and its surrounding watersheds. During Cousteau's negotiations in South America, he had gotten permission to enter the waters and territory of the host countries by promising that scientists from those nations would accompany his expedition. Cousteau told them that his films would clearly tie the watersheds of the world to the oceans, making a monumental ecological statement. More than three-quarters of the earth's population live within 10 miles of a coast or a major river. If the waters on which their lives depended were being polluted, they had to understand the relationship of the rivers and ocean to their survival in order to reverse course and avert disaster.

The Amazon was the perfect river with which to make his point. From the headwaters of its most distant tributary in the Andes Mountains of Peru, the Amazon and its countless branches collect more than 20 percent of the freshwater on earth and carry it to the Atlantic, 4,000 miles east. From the Amazon's 200-mile-wide mouth, the muddy plume of the river is visible from space, penetrating 300 miles into the ocean and filling the sea to a depth of 6,000 feet. During the dry season, 100,000 square miles in the river basin are covered with water; during the wet season, 300,000 square miles are submerged. More than a third of the planet's trees form the rain forest of the Amazon basin, which has the most diverse ecosystem of plants and animals on earth. More kinds of fish live in the Amazon than in the entire Atlantic Ocean. Until the middle of the twentieth century, the human population was limited to a few thousand widely separated bands living apart from one another and the rest of the world. In just fifty years, however, timber harvesting, large-scale agribusiness, and the resulting urban migration had thrown the once perfectly tuned ecosystem off kilter. No one other than scientists, a few hardy tourists, and the people of the Amazon themselves knew what was happening to the Brazilian rain forest.

Cousteau's expedition left Norfolk aboard *Calypso,* followed by a freighter headed for Lima and airliners bound for Quito, Ecuador, and Manaus, Brazil, which were the logistical centers supporting a half-dozen separate film crews. By the time the expedition returned to Norfolk a year and a half later, everyone had had enough of the Amazon. Falco had almost been stung to death by a swarm of bees. Jean-Michel, Anne-Marie, and ten others had gotten malaria. Cousteau had been severely bitten by fire ants on both arms, and they had been threatened by Colombian terrorists who believed that *Calypso,* equipped with a satellite dome and bristling with antennas, was a spy ship working for the CIA. But their film and videotape would produce seven of the finest hours of television to carry the Cousteau name and illuminate the plight of a river suffering under the burdens of a booming human population, ignorance, and abuse.

Under direction from Cousteau, who was rarely aboard *Calypso* or in the field with his camera crews, the river teams focused on collecting samples to gauge pollution and photographing fish and other wildlife. Jean-Michel and his crews, however, gravitated toward recording the lives of the people they encountered. Jean-Michel's selection of material was radically different from his father's—he believed that their films had to carry hard-hitting investigative journalism as well as beautiful shots of animals. He stopped for ten days to film the frenzy of forty thousand men in an open-pit gold mine at the base of the Andes. The 200-foot excavation was producing tons of gold but was also spewing tailings into streams and mercury vapor from its refineries into the air, where it mingled with water droplets that fell as rain hundreds of miles downriver. He found and filmed remnants of once isolated bands of people. Jean-Michel and his camera crews, who arrived by a float plane, were the first outsiders they had ever seen. Convinced that the cocaine trade on the Bolivian border was part of what his father called "the internal pollution of man" and contributed to the deterioration in the health of the Amazon watershed, Jean-Michel chewed coca leaves with farmers to sample the power of the drug, photographed vast coca plantations, and filmed the torment of a cocaine addict's withdrawal in a hospital in Lima.

As an unintended consequence of Jean-Michel's forays into social inquiry and human interest, in Peru he fell under the influence of a Jivaro Achuara Indian, who became the only mentor he had ever had other than his father. One of Jean-Michel's advance teams, led by an anthropologist who had lived among the Jivaros, introduced him to their charismatic leader, a short man in his late fifties known only as Kukus. Like many explorers before him who had encountered profound wisdom in a person regarded as primitive by the rest of the world, Jean-Michel found a deep connection with this man in the Amazon jungle. Kukus wore Western clothes, had seven wives, and had emerged as a powerful opponent of outside timber, fishing, and petroleum interests. They killed his people directly by bringing diseases for which they had no immunity, Kukus told Jean-Michel. They were also killing children who had not yet been born by taking away the forest and the river. For three weeks, Jean-Michel followed Kukus around, filming him as he planted trees, worked on a fish farm in a lake, and built a canoe. "I learned more about leadership from this one man than I have learned in my forty-six years of life elsewhere," Jean-Michel said. "From him I learned about the connectedness of everything."

Ted Turner was delighted when Cousteau and Jean-Michel brought back enough film for seven hour-long Amazon specials. He watched reels of raw footage and listened to tales of harrowing adventure, great beauty, and profound ecological insights during story conferences with Cousteau, Jean-Michel, and their writers. Editing the shows in Paris was going to take at least two years, so the first of them wouldn't air until 1985. In the meantime, Cousteau told Turner that he wanted to keep *Calypso* and his camera crews moving at full speed. Turner said fine. He had acquired exclusive television rights to the work of a man he considered to be an international treasure. Cousteau was one of his heroes and an expensive property but worth the money even if he was only breaking even on the deal. In the summer of 1983, Turner wrote a check for $2 million for an expedition to a less difficult but equally magical river, the Mississippi.

Calypso was in surprisingly good shape when she tied up in Norfolk after the Amazon expedition, so Cousteau was able to sail for New Orleans a month later. Jean-Michel, in charge of the day-to-day operations, was confident that if anything major went wrong on the river there were plenty of shipyards that would make room for *Calypso* on short notice. The plan was simply to spend a year sailing up the 2,300 miles of the navigable Mississippi, travel overland to film the headwaters in northern Minnesota's Itasca State Park, and detour into its major tributary, the Missouri River, for another thousand miles.

As always, Cousteau claimed that he would perform a scientific mission on the voyage, announcing that the society's science director, marine ecologist Richard Murphy, would take samples to measure pollution along the entire length of the river. The Mississippi was already the most studied major river on earth, so there was really no pressure to add to the enormous mass of data. Murphy spent just three days with his drogues and sampling tubes collecting less than a dozen samples.

Because the whole river system lived up to its nickname, Big Muddy, there wasn't much opportunity for underwater photography, either. Jean-Michel, with his father's permission, focused on the lives of the river's human inhabitants, as he had on the Amazon. More than anything, *Calypso*'s voyage up and down the Mississippi was a grand public relations tour orchestrated by the Cousteau Society and Ted Turner to return Cousteau to prominence with a new generation of television viewers. New Orleans, Memphis, St. Louis, and Minneapolis threw huge parties to greet *Calypso*. Cousteau sailed only a few miles of the trip on his ship but flew in for his celebrations and shooting scenes. Carefully distributed press releases alerted smaller towns along the way to *Calypso*'s passage. Crowds of people lined the banks and levies as it steamed by, some of them with bands playing "Aye Calypso." In Hannibal, Missouri, Cousteau played Mark Twain's organ. At one of Lewis and Clark's campsites on the Missouri, he entertained locals and the cameras with his bandonion squeeze box. In Minnesota, Jean-Michel cuddled black bear cubs at a research station, and one of the divers tried and hilariously failed to master logrolling at a lumberjack festival.

"The Mississippi River is not dead," Cousteau declared in his coda to the two-part television series. "Rather, it is vital, in all the senses of that word, reflecting the power and diversity of the culture along its banks."

Generalizations like this one did nothing for Cousteau's reputation among scientists, but at that point in his life, it only mattered that he made them. What Jacques Cousteau said on television was important to the world mostly because it was Jacques Cousteau saying it, and the truth was that oversimplifications and generalities were about as much as the average American could handle about ecology.

As he passed into his seventies, Cousteau had entered a realm of celebrity granted usually to beloved kings, presidents, and occasionally to athletes and movie stars. His name appeared on lists of the top ten most recognizable people on earth. He received thousands of requests every year to lecture, accept an award, or participate in a top-level international environmental gathering. France had made him a Grand Officier de l'Ordre National du Mérite and Commandeur de la Légion d'honneur, and awarded him every possible honor the nation could bestow. Monaco had inducted him into l'Ordre de Saint-Charles and made him a Commandeur ten years later. Belgium made him a member of l'Ordre de Leopold II, and Italy gave him its Ordine al Merito della Repubblica Italiana. He had been given gold medals and prizes by universities and exploration societies in ten countries, including the Gold Medal of the National Geographic Society, the United Nations International Environmental Prize, and membership in the American Academy of Sciences. Harvard, the University of California, Rensselaer Polytechnic Institute, and the University of Ghent had granted him honorary doctorates. When Cousteau was seated on the stage at the commencement ceremonies at the University of California at Berkeley, one of the graduating seniors detoured from the prescribed path after the president awarded him his baccalaureate degree to ask Cousteau to autograph his diploma.

While Jean-Michel was keeping tabs on the society and the Mississippi expedition, Cousteau spent most of his time in Paris or La Rochelle, where his second Turbosail ship was taking shape. Pechiney and the French government maintained their support after the *Moulin à Vent* disaster because Cousteau convinced them that the real problem with the ill-fated catamaran was that it had not been built from the keel up as a windship. With some of the money from his deals with Ted Turner for the *Odyssey* reruns and the shows on the Amazon and Mississippi, he had just enough to finish his second windship. In the summer of 1985, Cousteau was ready to make good on his promise to demonstrate the revolutionary propulsion concept by sailing it across the Atlantic and making a dramatic entrance into New York Harbor.

Cousteau named his new ship *Alcyone* after the daughter of Aeolus, the Greek god of the winds. It was built of aluminum, almost twice the size of *Moulin à Vent,* 103 feet long with the wide beam of an offshore racing sailboat and a monohull bow that fared back into a catamaran stern for stability. Two 33-foot-high Turbosails, mounted on the foredeck and amidships, propelled the ship at 10 to 12 knots in a steady 25-knot crosswind. In calmer conditions, a computer-controlled diesel took over automatically to maintain speed, using only 60 to 70 percent of the fuel normally burned by a ship of *Alcyone*'s size on an Atlantic crossing.

The crossing of the Atlantic this time was without incident, though the crew quickly realized that *Alcyone* was a very uncomfortable ride because it was built so stiffly. Even in a light chop, it lurched and bucked; every one of its twelve-member crew got seasick. Its entrance into New York Harbor on June 17, 1985, was a well-choreographed celebration laid on by the Cousteau Society, Ted Turner, and the city of New York. With fireworks bursting overhead and camera crews in helicopters circling, *Alcyone* and *Calypso* sailed past the Statue of Liberty. Cousteau had boarded *Alcyone* from the pilot boat as the windship cleared Sandy Hook. Simone was aboard *Calypso.* Two reporters from Turner's new international news network, CNN, fed live reports from the two ships. Every other news organization in America covered the arrival. Mayor Edward I. Koch of New York was on the dock to greet Cousteau and issue an official proclamation promoting him from captain to admiral. Koch then turned the microphone over to Admiral Cousteau.

"You will understand how moved I am to be received here in such a way with my old faithful ship *Calypso* and my new blond baby *Alcyone*," he told a crowd estimated at more than ten thousand. He told them of the challenges he had encountered in building a windship and sailing it across the ocean. "Ships are like women," he said. "Difficult to understand, but when you succeed it's worthwhile."

After two days of revelry and fund-raising in New York, Cousteau, Simone, Falco, Raymond Coll, and a select crew of current and former old hands sailed *Calypso* south to Chesapeake Bay and up the Potomac River to Washington, D.C. At the White House, President Ronald Reagan presented Cousteau with the Medal of Freedom, America's highest civilian decoration, for having done more than any other human being to reveal the mysteries of the oceans. Among the other recipients of the medal that year were Mother Teresa, Frank Sinatra, Count Basie, and the first man to fly faster than sound, Chuck Yeager.

The following day, across the river in Mount Vernon with eight hundred guests under an enormous white tent, Cousteau celebrated his seventy-fifth birthday a few days late. Three crews patrolled the tent with handheld video cameras, mics, and lights. Standing in front of Cousteau, who was seated in a comfortable wicker chair, John Denver sang a sweet rendition of "Aye Calypso," never breaking eye contact with the man he considered to be the greatest environmental hero in history. Ted Turner, looking like he'd arrived directly from a yacht club party in a casual summer suit, toasted Cousteau. "Happy birthday," Turner said. "God bless you, Captain Cousteau."

One of Cousteau's families was there: Simone; Jean-Michel, Anne-Marie, and their children, Fabien and Céline; Philippe's widow, Janice, and their children, Alexandra and Philippe Pierre Jacques-Yves; and Falco, Coll, and Papa Flash. When Perry Miller leaned down to embrace Cousteau in his chair, she knew they were both thinking the same thing: Philippe. She had been Cousteau's most energetic promoter and confidant at the beginning of his television career in America and knew how deeply connected he had been to his younger son. Cousteau seemed old for the first time.

"A birthday is entirely artificial," Cousteau said, typically banishing sentiment. "Nature doesn't count days. Monkeys and mosquitoes don't have birthdays." He was looking forward to his next seventy-five years, but even if he did not live to enjoy that birthday he was content. The Cousteau Society would continue his work long after he was gone.

Three months later, after every network and newspaper in the United States and Europe covered the birthday party at Mount Vernon, WTBS and CNN broadcast the hour-long *Cousteau: The First Seventy-five Years*. It drew the highest ratings ever for a Jacques Cousteau special on the superstation.

The Amazon series began to air shortly after the birthday party. Turner was sure his instincts had not failed him when he had decided to back Cousteau four years earlier, so he launched a massive publicity campaign. In magazine ads and endless promotional spots on all his radio and television stations, he touted the Amazon expedition as the most ambitious in history, the greatest and most difficult ever undertaken by Jacques Cousteau, the expedition of the century. The publicity surrounding the party in Mount Vernon had helped, too. Audiences soared when the first of the shows aired, with thousands of new subscribers paying to join his cable network. Turner was happy about the new business, but he also believed that Americans were living in ignorance of some harsh ecological truths and he had an obligation to bring Cousteau into their living rooms.

"If there is a mother of the environmental movement it was Rachel Carson," Turner said. "If there is a father, it is Jacques Cousteau."

Though the television audiences for the series were nowhere near what Cousteau had once drawn, many critics still applauded.

"One expects a Cousteau documentary to be beautifully photographed, despite the dangers and technical challenges of filming in primitive terrain," wrote the television critic of the *Hollywood Reporter*. "*Journey to a Thousand Rivers* exceeds expectation. Almost every frame of this production is exquisite."

Reviews of Jean-Michel's less traditional approach to making a Cousteau documentary were generally good, though some writers thought his focus on social issues was misplaced and overdone. *People* magazine said "Snowstorm in the Jungle," narrated by Orson Welles, had "the tone of *Reefer Madness* and *High School Confidential,* becoming almost more camp than compelling. Jacques Cousteau's comment in the film that 'the Western world may decline if the war on cocaine is not won' seemed a bit much."

When the shows on the Mississippi aired, scientists and environmentalists agreed that they were charming and entertaining but they didn't address a single important issue about the health of the Mississippi or its prospects for the future.

"Cousteau is to oceanography what cancan is to *Swan Lake,*" a French oceanographer told the Paris newspaper *L'Express.*

"Jean-Michel seems like a nice man," wrote another critic. "But he'll never be his father."

"Mississippi—Reluctant Ally and Friendly Foe" won Cousteau his first Emmy in a decade.

21

REDISCOVERING THE WORLD

My favorite ocean is the one I haven't been to yet.

Jacques Cousteau

AFTER THE AMAZON, the Mississippi, and the seventy-fifth birthday shows on TBS, Cousteau went to Atlanta to ask Ted Turner to finance a new series. Though the audiences on the cable channel were nowhere near as large as those on ABC or even on PBS, Turner had proven himself to be a publicity genius who made the most of his relationship with the world's most famous explorer. Cousteau, with Jean-Michel as his second, arrived early and waited in an empty conference room. A half hour later, Turner finally rushed in with an entourage of production executives, greeted the Cousteaus warmly, sat down at the head of the table, and closed his eyes. In a minute, he looked like he was asleep.

Cousteau glanced uneasily at Jean-Michel, shrugged, and started talking. On a five-year expedition, *Calypso* and *Alcyone* would retrace the routes of the great European ocean explorers, Balboa, Columbus, Cortés, Ponce de Leon, Magellan, and Vespucci. Cousteau would send his ships first to the Caribbean, then separately through the Panama Canal and around Cape Horn into the Pacific, across to New Zealand and Australia, and up into the great rivers that flow from the belly of Asia. His mission would be to document the changes in the pristine islands and continents that had been caused by human beings since the days of the great explorers. There would be plenty of adventurous underwater footage, but documenting the human interaction with the earth's rivers and oceans was the most important work they could do. What was runoff from farming doing to the Caribbean Sea? What was the waste from Australian cattle ranches doing to the Great

The windship Alcyone (AGENCE FRANCE-PRESSE)

Barrier Reef? What were the rivers of China, India, and Indochina carrying into the oceans? The list was endless. He called the series *Jacques Cousteau's Rediscovery of the World.*

"The *Rediscovery* series will have little to do with the behavior of animals," Cousteau said. "It will have to do with the behavior of people with respect to the water system. We will take a fresh look at the planet man believes he already knows."

Turner looked like he was still sleeping.

Cousteau went on, not sure what else to do. From the five-year voyages of *Calypso* and *Alcyone,* he would produce four finished hours of television each year beginning with the 1986 season. He would direct two of them, Jean-Michel would direct the other two. It was going to be expensive, Cousteau said, not wanting to mislead Turner. He had a second ship and *Calypso* was probably going to need new engines before challenging the Pacific. He wanted to give his divers and crewmen an updated image with new technologies in underwater equipment, electronics, and cameras, a lesson he had learned from David Wolper. They would be fools to think that a new generation of television viewers was going to respond to the same thing their parents had when they watched *The Undersea World* and the *Odyssey* series. Each episode would cost $1 million.

For a long minute after Cousteau stopped talking, the room was silent. Then, as though startled, Turner leaned forward in his chair and said, "Let's do it."

When they worked out the details, Turner agreed to pay Cousteau $750,000 per episode in 1986, escalating to $950,000 for each of the four hours delivered in 1991. He paid an advance of $3 million for the first year right away, with similar advances to come after he got the fourth show of each season. Cousteau had complete control over the content of his documentaries. Turner had the rights to broadcast each new episode during a prime evening hour, and as many times as he liked as a rerun. Each hour was going to cost $1 million to $1.2 million to produce, so even with Turner's backing, the Cousteau Society had to raise an additional $250,000 per episode.

Jean-Michel would begin to gradually replace his father as the star of the series, appearing more frequently aboard both ships and taking center stage in the narration. By the end of the five-year contract with Turner, Jean-Michel and his father wanted to be interchangeable in the minds of their audiences to ensure the future success of the Cousteau Society and its television production company. They were infinitely practical about the realities of Cousteau's age and inevitable death and the costs of keeping their film crews at sea. Cousteau would spend most of his time raising money.

Though *Alcyone* had proved the fuel-saving concept of the Turbosail, the royalties Cousteau expected it to generate never materialized. Oil was cheap again, so shipping companies were no longer interested in spending money to refit their freighters and tankers with wind power. Cousteau continued to believe that the Turbosail would become irresistible as soon as the price of oil climbed, as it was bound to. For the time being, he needed a new idea to help finance his television shows.

As they prepared for the *Rediscovery of the World* expeditions, Cousteau and Jean-Michel launched a venture they hoped would provide a steady stream of cash for film production to take pressure off the society. After the success of the oceanographic exhibits in Monaco and aboard *Queen Mary* in Long Beach, California, Cousteau was convinced that people would come from hundreds of miles around to see the wonders of the deep in an oceanographic park similar to Disneyland. With Jean-Michel in charge of the project, the Cousteau Society loaned Parc Océanique Cousteau $1 million. It was enough to build the park, but it would have to break even or make a profit immediately to survive beyond its first year of life. Jean-Michel was optimistic about his chances for success because he believed the parc would become the main attraction of the renovated Les Halles district in a section of Paris that had been plagued by urban blight. The hope was that shiny stores, restaurants, and amusements would revitalize it. If the concept of introducing people to the wonders of the ocean in a downtown theme park worked, Cousteau ocean centers would follow in Los Angeles, Norfolk, Brazil, and Japan.

As *Flying Calypso* had been Cousteau's paternal gesture to Philippe, the Parc Océanique affirmed his love for Jean-Michel. Cousteau had known instinctively that Philippe needed to succeed in some way apart from the glare of his own charisma if he were to continue as a member of his circle of intimates, on whom the future of his work depended. It was the same with Jean-Michel. Always a builder and a student of architecture, he had hungered to eclipse his famous father in at least one endeavor. Cousteau, like everyone else, could see that Jean-Michel was always going to be a less-than-stellar replacement when he died. Jean-Michel made documentaries simply because he had to and because his skill as a natural organizer transferred effectively to film production. Cousteau believed the Parc Océanique had a chance to make money, but if it succeeded, he also would have done his full duty as a father.

Under Jean-Michel's direction, the exhibit designers who created the Pirate's Cove exhibit at Disneyland began building Parc Océanique Cousteau using the most interactive, sensational technology ever developed for an amusement park. When it opened in the summer of 1988, visitors would plunge from a rocket in space into the ocean depths, where animals seen only from research submarines came to life in the strange light of the abyss. If people had questions, they could ask them of a life-size replica of Jacques Cousteau, which would answer them in the familiar voice from their television sets. They would tour the inside of a full-size blue whale, touch its beating heart and other organs, and live to tell about it. On a jungle gym covered with carpeted sharks, children would overcome their fear of the ocean's most frightening creatures. In a theater with a giant 45-foot screen, audiences would experience the films of Jacques Cousteau as never before.

At the end of August 1985, *Calypso* and *Alcyone* sailed from Norfolk to rediscover the world. Haiti was the first stop, chosen simply because Cousteau had never been there before. He found a country of six million people packed into a third of the tiny island of Hispaniola, also home to the Dominican Republic. More densely populated than India, Haiti was in the midst of a catastrophic environmental disaster, an example of what Cousteau feared would happen to the rest of the world if people did not wake from their slumber of complacence about the places where they live. He toured the island with his own handheld video camera, creating a sociological portrait of desperate people somehow living optimistically in the ruins of their fields, watersheds, and forests. Haitians suffered from raging epidemics of diseases caused by pollution and contamination of their resources. One in ten infants died at birth. Cousteau's own crewmen fell ill from eating toxic fish. In his narration of the episode shot in Haiti, Cousteau celebrated the brave people of the crippled nation, "the spirit of the Haitians themselves, who, while facing a troubling future, endow the present with an inviolable human grace." On the day before *Calypso* and *Alcyone* sailed away, Cousteau visited Haiti's most inspiring shrine, a waterfall, where, according to legend, the Virgin Mary had appeared fifty years earlier. He joined pilgrims under the sacred waters, where, as one of his writers noted, "celebrity and celebrant were the same, as each was refreshed and renewed in a rite celebrating the source of life—water." Cousteau promised to return to help Haitians revitalize their streams, bays, and shoreline with sea farms. His title for the Haiti episode was "Waters of Sorrow."

Both ships then called at Havana, where Fidel Castro received Cousteau like a visiting head of state, listening attentively as Cousteau lectured him about the desperate need for the end of human rights abuses and appealed for the release of Cuba's political prisoners. In Cousteau's honor and with no fanfare, Castro freed fifty prisoners shortly after *Calypso* and *Alcyone* sailed away. One of the prisoners, a former art professor named Lázaro Jordana, broke the news about Cousteau's amnesty only after he had fled the country for a new life in Paris.

"Cousteau saved my life, and my father's life, and the lives of all of the fifty people released," Jordana said. "Then he didn't say anything about it. No publicity. He's the kind of man who does things—doesn't talk about them—just does things."

Three months later, *Alcyone,* under the command of her captain Bernard Deguy and with Jean-Michel aboard, sailed for Cape Horn while *Calypso* headed for a shipyard in Miami. Even after an overhaul three years earlier, her forty-five-year-old engines were shot. Their blocks, cylinders, and clattering bearings produced barely enough power to run at half speed. Cousteau didn't want to risk beginning an extended cruise around the Pacific without replacing them. He wrote a letter to the members of the Cousteau Society explaining *Calypso*'s plight and asking for a special donation to pay for new engines. They cost about $160,000, including propellers, shafts, gearboxes, and spare parts. A month after he wrote the letter, checks totaling $260,000 had arrived at the society's office in New York.

In July 1986, *Calypso* was ready for the sea. After another gala send-off, with what Cousteau called "the courage of new engines," she transited the Panama Canal and struck out across the Pacific for New Zealand and Australia. *Calypso*'s arrivals in Auckland, Melbourne, Sydney, and other major and minor ports were causes for celebration. The following summer, *Calypso* headed north via Papua New Guinea for a rendezvous with *Alcyone.* While *Calypso* was in the Miami shipyard, *Alcyone* had rounded Cape Horn in calm weather and worked her way up the coasts of South and North America, filming targets of opportunity on the way to Alaska and into the Bering Strait between Russia and North America. Every port call along the way was a chance to show off the Turbosail, make newspaper headlines and television broadcasts, and keep Cousteau and his adventures alive for millions of people.

The routine aboard both ships was a steady grind of production deadlines and the uncertainties of life at sea. For the twenty-six men and Simone aboard *Calypso,* and the eleven men aboard *Alcyone,* the voyages from one shooting site to the next were scrambles to repair equipment and figure out what might or might not fit into a particular television hour. *Alcyone* was still a very hard ride in any weather, and *Calypso* still rolled when the wind blew. They shot without scripts, capturing whatever was interesting, both on land and in the water. They shipped videotape back to New York or Paris and picked up dispatches from editors about the progress of a particular story line with requests for footage to fill holes.

With a few exceptions, the crews were French, the cameramen mostly Parisians, the divers and seamen from Brittany and Marseille. When Cousteau or Jean-Michel were not aboard, Falco and *La Bergère* were in charge of *Calypso.* As had been the custom since Calypso's first expedition to the Red Sea in 1951, everyone aboard each ship had more than one job except for the cook. Everyone stood wheel watches. At an underwater shooting location, diving took three hours of the day, preparation and repairs another four hours, leaving another seventeen hours for sleep and socializing. Getting along aboard small ships was crucial to success. If a man failed as a pleasant companion, he wasn't asked to stay aboard for the next leg of the voyage. The decision was made by vote of the crew. Being asked to remain was an honor.

Once his ships were in the Pacific, Cousteau spent most of his time traveling to raise money and give speeches, or in Paris supervising his editing studio and tending to his second family with Francine. In 1988, he resigned as the director of the Oceanographic Museum of Monaco. Though he continued to speak on its behalf at conferences and celebrations, he spent very little time on the Riviera. He was also far less interested in the daily grind of television production on location, much preferring the comforts of home or a good hotel. Cousteau was in constant demand as a lecturer. He sometimes appeared in cities on three continents in a single month to deliver his message, which had evolved from sounding the alarm about the deterioration of oceans and rivers into appeals for world unity, population control, and the abolition of nuclear weapons and power plants.

The key to balancing human needs with available energy resources, he said, was the abolition of waste and the development of solar power. "Look at the English," Cousteau told an audience at an international energy conference. "Its wealth was coal, but coal is finished. The United States has oil, and its reserves are running down. We're drawing down everything on the planet. One day only solar energy will be seen as truly inexhaustible and sun-drenched, semi-tropical, and tropical countries will be rich."

Cousteau spoke to the Association of Space Explorers, an elite club of astronauts and cosmonauts who had flown in space. "Space explorers and divers all seem to share the belief that disputes among nations are vestiges of a less enlightened time when humanity did not clearly understand that all people were dependent upon one another for survival," he told them. "We must solve our conflicts once and for all so we can begin our new adventure of exploring the universe."

Later, Cousteau proposed a way to remove the threat of nuclear war that he devised with futurist and science fiction writer Arthur C. Clarke. Nations facing one another in a nuclear standoff, such as the United States and the Soviet Union, would simply agree to exchange all their children for at least a year. "Imagine a world where all the children from seven to eight, for example, would have to spend a year on the other side of the fence," Cousteau said. "From an educational standpoint it would be a great opening of the mind. It is because of the hope that I have in the future of mankind that I want to join forces and forge a better world for future generations."

From New York, Norfolk, and Paris, Jean-Michel orchestrated the production of four hours of television a year for five years, rarely spending more than two or three days in one place. His father spent perhaps a weekend a month aboard one of the ships, usually arriving late, disturbing the routine, walking through his close-ups, and leaving. Jean-Michel told his camera crews to use back shots and doubles for Cousteau if they absolutely had to have him in the frame and he wasn't on location. Otherwise, he edited in sequences from other locations when his father had been aboard or on the scene of a similar shot on land. Because the demand for convincing doubles was high, the job of a crewman who from behind looked like Cousteau in a red knit watch cap was among the most secure on both ships.

When Jean-Michel wasn't with his film crews on the Pacific, he was in Paris coaxing the Parc Océanique to life. Six months after it opened, however, it was obvious that the venture would fail. L'Équipe Cousteau and the promoters of the Forum des Halles development launched it with a brief publicity campaign. Newspapers and television stations reported on the wonderful new way to learn about the ocean without an aquarium or any live animals at all. But after the grand opening, the crowds fell to a trickle. Some days only a few dozen visitors bought tickets. The critical mistake, the Cousteaus quickly figured out, was to have built the Parc Océanique in an underground mall that was part of a massive downtown development project. Not many people, it turned out, wanted to tour the ocean at a subterranean amusement park with no live animals. By the autumn of 1990, with the initial capital from the Cousteau Society running out and no chance of other investors, Jean-Michel and his father began an orderly retreat into bankruptcy. Parc Océanique was a separately incorporated venture, so the Cousteau Society could lose only its initial investment with the default on the million-dollar loan. For reasons Jean-Michel could not fathom at the time, his father's generosity in encouraging him to build the Parc Océanique turned bitter, as Cousteau allowed the failure to rest squarely on his son's shoulders.

With two ships, crews, and camera teams at sea all the time, Cousteau was barely paying the bills. There was no chance that the *Rediscovery* series was ever going to be worth any more than the society was already getting. Turner had been right about cable television revolutionizing the industry, but the unintended consequence of offering viewers hundreds of channels twenty-four hours a day was a watering down of audiences for all but the most sensational sporting events and other unique offerings. *Jacques Cousteau's Rediscovery of the World* had a built-in following of society members who ritualistically tuned in once a quarter to see what their money was buying. Entire cable channels were dedicated to airing natural history and sociological documentaries, which competed with the adventures of the *Calypso* and *Alcyone* in the Pacific. Turner was known for unsentimental financial decisions, so even his closest advisers and executives were puzzled when he extended Cousteau's contract for three more years, until 1992. Then, as though to contradict himself, he downgraded the importance of the Cousteau series by placing a junior producer in charge. Cousteau was furious, at first refusing to even acknowledge the presence of the young producer, Tom Beers, at meetings. Beers quickly saw that Cousteau was not at all involved in the day-to-day details of production and that Jean-Michel, with whom he got along well, was really in charge, so he waited out the conflict and survived.

Cousteau was also less than fully engaged in the life and camaraderie of *Calypso* and its crew, which had sustained him for four decades, withdrawing most of the time to his apartment in Paris. He continued to make public appearances, accepting awards and honorary degrees, speaking on behalf of the society, and attending conferences, but he saw few people outside his immediate families. Dumas was dying. Tailliez was rarely in touch from his villa near Toulon, where he presided over the memory of *Les Mousquemers* and the beginnings of scuba diving. Laban, who had pulled away from Cousteau after the death of Philippe, spent most of his time painting at home on the Mediterranean coast. Cousteau knew thousands of people, millions claimed to know him, but intimate moments of friendship did not exist in the glare of celebrity, except in his secret life with his mistress.

Cousteau and Simone had led distinctly separate lives since Philippe died. They were friendly during their encounters aboard *Calypso,* at family gatherings in Paris or Monaco, or when they were together at award ceremonies or formal occasions. Otherwise, Cousteau's life had flowed into that of Francine and their children. Simone rarely left *Calypso.* In early 1990, she went ashore to see a doctor for a checkup, found out that she had an aggressive cancer, and went back to her ship to die.

On December 3, 1990, Jean-Michel flew into Bangkok to rendezvous with *Calypso* and called his office in Paris from an airport pay phone. The day was already tinged with melancholy because the Mekong River expedition was going to be the last for *Calypso.* After fifty years at sea, she was constantly breaking down and the repair bills were endless. A week before, his father had ordered it to be tied up after the shooting in Southeast Asia until he decided how to bring it back to France for good. From then on, they would use *Alcyone* and begin planning for the construction of a new ship, *Calypso II.*

When Jean-Michel finally got through to Paris, there was only one message that mattered. *La Bergère* was gone. She had died after a mercifully brief encounter with her disease. His assistant in Paris told him that Cousteau had reached Simone's bedside in time for a farewell.

Later that week, Cousteau and his first family gathered in Monaco. On a bright afternoon for December, they boarded a launch owned by the Oceanographic Museum and motored slowly from the harbor. On the headlands, dark clusters of people stood silently as the funeral passed beneath them. Two miles offshore, after the ceremony of military honors conducted by a detachment from the French navy, the ashes of Simone Melchior Cousteau became part of the winter-blue Mediterranean Sea.

22

CHAOS

JEAN-MICHEL STAYED IN FRANCE for a while after Simone died. He wanted to be near his father, even though Cousteau was keeping his distance from everyone, including his son. Shortly after the new year, Cousteau asked Jean-Michel to lunch. They had eaten countless meals together but never one for which his father had extended so stiff and formal an invitation. I have something important to tell you, Cousteau said. At the cafe, Cousteau wasted no time after the waiter left their table to turn the world upside down. Cousteau said that he and Simone had not lived as man and wife for a long time before her death. For fifteen years, he admitted, he had been with another woman, Francine Triplet, whom he intended now to marry. They had two children together, both of them, by then, teenagers. Jean-Michel's stepsister was Diane, his stepbrother Pierre-Yves. Cousteau said he had kept his life with Francine secret out of respect for Simone, but it could no longer remain a secret.

Jean-Michel had known his father loved women. It was part of his charm, one of the reasons women loved him. He also knew that his mother's separate life aboard *Calypso* was a practical solution for maintaining appearances and above all not damaging her husband's reputation. She was, after all, a navy wife, the daughter and grand-daughter of French admirals. Appearances mattered. Jean-Michel had no quarrel with his father's insatiable appetite for women, but he never thought one of them would replace his mother. The news that Cousteau had a second family and a deeply concealed secret life was devastating to Jean-Michel for one reason alone: Did his mother know? Cousteau told him he wasn't sure. Possibly.

After that day, Jean-Michel lost himself in his work. He had to get *Calypso* out of Southeast Asia. The Parc Océanique was collapsing,

Jacques and Francine Cousteau
(AGENCE FRANCE-PRESSE)

and now he knew that it was probably Francine who had led his father to blame him for the failure. He had to deliver two episodes to Ted Turner right away. Then suddenly on June 28, 1991, none of it mattered. Just six months and a few days after Simone had died, his father and Francine Triplet were married in a quiet ceremony. Whether by design or coincidence, the day was the twelfth anniversary of Philippe's death.

Since Cousteau revealed the truth about his second family, Jean-Michel had been tormented by thoughts of his mother in her self-imposed exile aboard *Calypso*. Had she known that the man with whom she had opened up the undersea world, with whom she had raised two sons, and with whom she had mourned the death of one of them, had replaced her with a younger woman? That his father had the audacity or the ignorance to marry his mistress on the very day his brother had been killed twelve years earlier was unforgivable. As soon as Cousteau surfaced after his wedding, Jean-Michel told him he was through. He would see the last two episodes of the second *Rediscovery* series to completion. Then he wanted nothing more to do with Cousteau, his new family, the society, or anything else that was still alive in the smoking embers of what had been a grand life. Cousteau embraced his son, said his affection for him would never diminish, but as he had with everyone who committed the ultimate act of betrayal by leaving him, he banished Jean-Michel into the unmentionable recesses of the past.

With Jean-Michel gone, everyone in business with Cousteau was shaking their heads at the grim prospects for the future. Publicly, they expressed optimism for the continuation of Cousteau's work, but most of them privately believed that all he had built might very well die with him. At the end of Francine Triplet's career as an Air France flight attendant, she had been a senior executive overseeing in-flight service. She participated in the making of several promotional films but had no other experience that would explain why L'Équipe Cousteau hired her as a scriptwriter later that year. It was even more puzzling that Cousteau seemed to be grooming his wife to take over after his death. Increasingly, the new Madame Cousteau was making decisions about the society and L'Équipe Cousteau on behalf of her husband.

In 1995 Cousteau turned eighty-five. His lungs were failing, and he had withdrawn from almost all activity outside his home in Paris. The man who taught the world how to breathe underwater, he joked, was having trouble breathing on land. He was equally cavalier about the prospect of his own death.

Cousteau still granted interviews, intermingling charm, philosophy, and the inevitable promotion of the society and his television programs. "If I have cancer, so what? That's a way to finish your life. It's one more sickness. It's nothing terrible. I mean, yes, it's terrible, but death is terrible in itself. I have made friends with death," Cousteau said. "I mean I have accepted it not only as inevitable but also as constructive. If we didn't die, we would not appreciate life as we do. So it's a constructive force."

He said he regretted only that it was impossible for him to continue to live as an animal when he was sick. "When an animal is hungry, he will hunt several weeks without sleep. When he has eaten he will sleep for three days. That's the way I go."

Cousteau held out no hope for an encounter with a benevolent and forgiving god after this life. "If there is a god and He's interested in life, He's just as interested in a French poodle as in you or me," he told one writer.

Why were you able to expand human awareness of the need to protect the environment, a reporter asked? "If *The Silent World* was such a success when it came out it was, of course, because people were beginning to have a feeling for their surroundings and a curiosity about the sea depth that no one had seen before. But it was also because I knew how to make a movie."

In another interview, he repeated his long-held conviction that humanity had to begin thinking about the long-term health of the planet or face the same destruction he had witnessed in Haiti.

> The earth is probably two-and-a-half billion years old, and we know that in about five billion years life will be impossible on earth because the sun is going to expand and burn everything. So we're about a third of the way in the life of the earth, which means that if we take care of it, humans can plan for several billion years on this wonderful planet. Things change, of course. The world is not what it was. We must plan long-term. Accordingly, to save it for our distant children, we must establish four priorities. The first one is peace. We know it is difficult, but there must be ways to live in peace other than leaving it in the hands of governments.
>
> The second priority is limiting our own number. Rich nations have stable populations, poor nations are a time bomb. The third priority is education. If we want to do something for peace and population, for the Third World and the environment, we have to demonstrate the problem. The fourth priority is the environment. If I let my reason speak I am not optimistic. I don't see any possibility to change people, the people who make the decisions. However, I believe that by action, faith, and hope we can achieve something. Each one of us must do something to fight for peace, for better cooperation among people, for education and the environment.

Among friends with a bottle of wine on the table, Cousteau's pronouncements were much less pained and more hopefully eloquent:

> I find poets closer to the truth than mathematicians or politicians. They have visions that are not only fantasy. They are visions that are, for some reason they cannot explain, an inspiration that guides them

and brings them by the hand, or by the pen, closer to the truth than anybody else. I believe we should follow the poets more than anybody else in life. It's the light. It's the star we should be guided by . . . The only remedies to the logical absurdities are utopias, reasonable utopias.

As Jacques Cousteau faded further into the background, his new wife emerged as the heir to all he had created. Falco, Raymond Coll, and the few original *Calypso* divers who were still around watched Cousteau's health decline and prepared to quit, as did many writers, editors, and other staff. They had been loyal to Cousteau and Jean-Michel but could not understand why he was leaving everything in Francine's obviously incapable hands. She seemed to know very little about running expeditions, managing nonprofit corporations, or making movies.

While Francine shored up her defenses against criticism and the abandonment by Cousteau's loyal friends, Jean-Michel struggled to establish a separate identity. His father was still bristling because his son had quit on him, and the bitterness took on a sharper edge because Francine was increasingly protective of her husband's privacy. It was hard for Jean-Michel to see his father even for purely personal reasons, and his misunderstandings with his stepmother deepened. Shut out and suffering from it, Jean-Michel turned his attention to lecturing and leading environmental tours and cruises. He went into business with a developer who was building a vacation village on Vanua Levu, one of the Fijian islands. The resort was among the first in a new wave of what developers were promoting as eco-friendly, a small collection of thatched huts around a pristine lagoon. Guests could scuba dive, snorkel, and learn about life on a primitive South Pacific island, but no fishing, motorboats, or Jet Skis were allowed. With Jean-Michel as a partner, Vanua Levu was billed as a Cousteau environmental resort. Despite his banishment from Cousteau's inner circle and the obvious management of his father's affairs by Francine, Jean-Michel was stunned when his father sued him to prevent him from using the Cousteau name.

After a brief exchange of legal threats, they settled out of court through their lawyers and without ever reconciling privately. Jean-

Michel agreed to use his full name—Jean-Michel Cousteau—for any of his own ventures to differentiate them from those of the Cousteau Society and his father's other enterprises. He told reporters that the theatrics of a public lawsuit were completely uncalled for.

"My father and I could have solved our differences over a good bottle of French wine and an embrace." Jean-Michel was certain their problem could have been solved because without Francine, JYC had never been a grudge holder. To hold people accountable for past transgressions would have violated his cherished code of living in the present. If someone betrayed or mortally offended Cousteau, they simply ceased to exist, something that was out of the question when family was involved. He had shown over and over that no matter how serious a breach or a rupture in his relationship with his sons, his heart had remained open to them. When Philippe tried and failed to go it on his own as a filmmaker, JYC took him back without hesitation. When Jean-Michel quit the Cousteau Society to protest the hiring of Fred Hyman as its director, he was still part of the family. Everything changed with Francine as part of the equation.

"JYC was not a superman," Jean-Michel insisted ten years after his father's death. "He was just an ordinary man who made a bad mistake. Living a secret life with a second family diminished him and darkened his outlook about himself and the earth as he grew old. By the time he died, he no longer believed that humanity could save itself from disaster. I disagree with him, but I do not hold his mistake against him. He was my father."

As though to foreshadow the end of Cousteau's life, *Calypso* sank in January 1996. After Simone's death, Jean-Michel had finished the Mekong River expedition, left the ship in Ho Chi Minh City for almost a year, then took it to Singapore. Falco was there trying to get *Calypso* in shape for a final voyage back to retirement in France when a drifting barge slammed into it. In five minutes, it was on the bottom in 16 feet of water, with its funnel, masts, and the crow's nest on its foredeck above the surface. A week later, Falco mustered a salvage crew, raised *Calypso,* and loaded it aboard a barge bound for Marseille. A few days before his ship made it home to France, Jacques Cousteau had a heart attack. He hung on for two days, fading in and out of consciousness, during which time Francine refused to allow Jean-Michel to see him. Jacques Cousteau died at two thirty on the morning of June 25, 1997.

Five days later, on a soggy Paris summer morning, the world sent Cousteau into eternity with a funeral mass at the Cathedral of Notre Dame. A French naval honor guard carried his coffin into the cathedral, followed by his wife, their two children, Jean-Michel, the rest of his surviving family, President Jacques Chirac, and a thousand other mourners. Outside, thousands more stood in the rain, straining to listen through the open stained-glass windows as the archbishop of Paris chanted the mass and, in his sermon, called Cousteau "a poet of an inaccessible reality." President Chirac, speaking on behalf of his nation, said Cousteau was "an enchanter who represented the defense of nature, modern adventure, and the dreamy part at the heart of all of us." Jean-Michel spoke next, with the strained tones of a heartbroken man struggling to be brave: "The work of my father was a hymn to life. On the wall of my office is a quotation from him: 'The happiness of the bee and the dolphin is to exist. For man it is to know about existence and to marvel at it.' "

The day after the funeral mass in Paris, Jacques Cousteau was buried in the cemetery at St.-André-de-Cubzac next to the tomb of his mother and father, facing northwest up the Gironde Estuary toward the ocean.

The Cousteau family tomb, St.-André-de-Cubzac, France
(COURTESY OF THE AUTHOR)

EPILOGUE

FOLLOWING COUSTEAU'S explicit wishes, the boards of directors of the Cousteau Society and L'Équipe Cousteau elected Francine Cousteau president of both organizations. Turner Broadcasting aired several tributes to her late husband, along with reruns of the *Odyssey* and *Rediscovery* series, but there were no Jacques Cousteau films in production for the first time in fifty years. Without the publicity from new adventures, membership in the society and L'Équipe Cousteau went into a steady decline, falling to about half its peak of three hundred thousand by the turn of the new century. Francine Cousteau had complete control over the film archives, records, photographs, publications, and the use of the family name. Jean-Michel, completely cut off from his father's estate after the suit over Vanua Levu, turned his attention to his tourism ventures, speaking engagements, and planning for his independent return to filmmaking. He founded his own nonprofit corporation—the Jean-Michel Cousteau Foundation—and led a campaign to return Keiko the killer whale to the wild after years of captivity and fame as the central character in the movie *Free Willy*. When Jan Cousteau, Philippe's widow, announced plans to establish a foundation using the Cousteau name, Francine successfully sued to stop her, banishing that branch of her late husband's family as well.

"They want to capitalize on the name," she said. "If their name was not Cousteau, nobody would know who they are. I have to put a stop to it."

"Just because her last name is Cousteau, Francine thinks she is a Cousteau," Jean-Michel said. "She will never be a Cousteau."

A bitter fight raged between Cousteau's two families over possession of *Calypso*. After Falco brought the ship back to France, it lan-

Jacques Cousteau, the Sea King
(AGENCE FRANCE-PRESSE)

guished for two years on a barge in Marseille until Francine, backed by a court order, moved it to a dock at the French Maritime Museum in La Rochelle, several hundred miles to the northwest. There, *Calypso* rotted and sank lower into the water while a five-year legal battle dragged on between Jean-Michel and Francine.

Years before, Cousteau had told Jean-Michel that he and Simone wanted *Calypso* taken out to sea and scuttled when it was no longer fit for service. Neither of them could bear the thought of tourists roaming around her decks spilling their Coca-Colas on the ship that had been as important to them as their own hearts. Jean-Michel, and later Francine, begged Cousteau to reconsider, but he held firm. After his death, neither Cousteau's new wife nor his son could sink the ship that had become so precious to France and the world as a symbol of exploration and enlightenment.

Jean-Michel; his children, Fabien and Céline; his niece Alexandra; Falco; Coll; and most of the original *Calypso* crewmen wanted the ship to become the centerpiece of a new oceanographic museum on the

Mediterranean. They believed they could raise money from Cousteau's millions of admirers to restore the ship and build the museum, which would be a public institution. Francine, meanwhile, made a deal with a cruise ship company to move *Calypso* to the Caribbean, where it would become an attraction for the company's passengers and other tourists. Jean-Michel quashed that plan with a countersuit based on French customs law, under which *Calypso* was considered an object rather than a working ship, and therefore could not be exported so easily.

"*Calypso* belongs in France," Jean-Michel said when he announced his victory. "It is a national treasure."

With no end in sight to the ownership dispute, L'Équipe Cousteau came up with the money to repair *Calypso*'s hull enough to keep it from sinking, cover the leaking decks with tarps, and pay for continued moorage in La Rochelle. A full restoration was going to cost $1 million, maybe $2 million. In the wake of the expenditure of hundreds of thousands of dollars in legal fees fighting over the family name and dramatic declines in membership, neither L'Équipe Cousteau nor the Cousteau Society had that kind of money.

Despite the suits and countersuits filed by Francine and Jean-Michel, the estate of the late Loel Guinness was still the only legal owner of *Calypso*. Guinness had leased rather than sold the ship to Cousteau fifty years earlier for the token sum of one British pound. The lease was still held by Cousteau's original nonprofit French Oceanographic Expeditions, whose board was made up of old *Calypso* hands loyal to Jean-Michel and the Mediterranean plan for the ship. As the head of L'Équipe Cousteau, Francine claimed a seat on the COF board, hoping to wrest control of the ship that way. To thwart her, the old hands canceled the meeting at which she was going to demand her place.

Finally, Carnival Cruise Lines offered L'Équipe Cousteau $1.3 million to rebuild *Calypso* and keep it under the French flag. Loel Guinness's grandson, convinced that the offer was the only hope for saving the ship, canceled the original lease to COF and sold it to Carnival for one euro. After a final round in court, a judge approved the deal.

In the fall of 2007, a pair of tugboats carefully towed the rust-streaked, wallowing hulk north along the Atlantic coast from La Rochelle to a shipyard in Brittany for resurrection. It was unrecognizable as the magical set for such scenes as the storm on its first voyage under Cousteau on the Mediterranean, when it proved strong enough to survive anything; or Cousteau's triumphant arrival in New York harbor for the World Oceanographic Congress; or its slow voyage up the Mississippi River with cheering crowds lining the levies and banks. It was impossible to connect the lifeless, dingy hulk to the thousands of days when divers left *Calypso* to explore the ocean, capture the underwater world on film, and bring it to televisions in hundreds of millions of homes, hard to imagine that this was the ship that had revolutionized human understanding about the sea and its creatures.

According to the Cousteau Society, *Calypso* will be ready for sea again in early 2010.

NOTES

Prelude: Autumn 1977

—Cousteau's fund-raising mission in the United States and the event in Seattle are described in contemporary newspaper accounts and confirmed in a November 2005 interview with an official of the Bullitt Foundation. The foundation made a donation of $3,500 to the Cousteau Society.

—Cousteau's preoccupations in the fall of 1977 are reconstructed from the timeline in Jacques-Yves Cousteau and Alexis Sivirine, *Jacques Cousteau's "Calypso"* (New York: Abrams, 1983), an account of the ship's voyages. According to that account, the pollution survey voyage began on July 27, 1977. In late October or early November, the crew of *Calypso* dove off the coast of Otranto, Italy, on the wreck of the Yugoslavian freighter *Cavtat,* which had sunk three years before while carrying 300 tons of tetraethyl and tetramethyl lead.

—Cousteau's interview with James E. Lalonde and Richard Strickland appeared in the November 2, 1977, edition of the *Seattle Weekly.*

—The details of the Mediterranean survey are from the film "Mediterranean: Cradle or Coffin?," an episode in *The Jacques Cousteau Odyssey* series, which first aired on PBS on May 27, 1978; from a Cousteau Society background paper on the expedition and the television production; and from a report on the expedition in the November/December 1977 edition of *Calypso Log.*

—Confirmation of Philippe Cousteau's presence in Los Angeles recovering from a leg injury suffered in the crash of his gyrocopter on Easter Island and the activities of Jean-Michel Cousteau are from an interview with Fabien Cousteau in the spring of 2006.

—The beginning of Cousteau's affair in 1977 with Francine Triplet, who would become his second wife in June 1991, is established in an interview Francine gave to *E: The Environmental Magazine* in June 1999, in which she states: "I spent twenty years of my life with him . . ." Cousteau died on June 25, 1997, which therefore dates their relationship to 1977. The date of their meeting may also be inferred from the ages of their two children (Diane, fourteen, and Pierre-Yves, twelve) at the time the relationship became public knowledge after the death of Simone Cousteau in December 1990. Additionally, Jean-Michel Cousteau, in an interview in April 2008, states that during the autumn of the Involvement Day campaign (1977), his father was reported to have attended the event in Houston while accompanied by Francine Triplet.

—That Cousteau was rarely aboard *Calypso* by 1977 is confirmed in interviews with former crewmen André Laban and Marc Blessington, and Fabien Cousteau.

—Cousteau's praise of Simone in her role as *La Bergère* is well known, but this specific statement at the end of the Prelude is taken from "Soul of the *Calypso*," by Samuel G. and Debborah Lecocq, which appeared on the Web site http://www .portagequarry.com in March 2006. Samuel G. Lecocq sailed with Jacques and Simone Cousteau aboard *Calypso* on several occasions.

1: *La Bergère*

—Cousteau's statement that marriage is archaic is from an interview by Sara Davidson for the *New York Times Magazine,* published September 10, 1972.

—The meeting of JYC and Simone in 1936 at her parents' apartment in Paris is confirmed in an interview with Fabien Cousteau, and in many published accounts, including Leslie Leaney, "Jacques-Yves Cousteau: The Pioneering Years," *Historical Diver* no. 13 (fall 1997).

—Descriptions of Simone Cousteau are drawn from early photographs.

—The details of Cousteau's automobile accident and injuries in 1936 are from Leaney, "Jacques-Yves Cousteau"; Axel Madsen, *Cousteau: An Unauthorized Biography* (New York: Beaufort Books, 1986), 13–20; and other sources.

—Descriptions of St.-André-de-Cubzac, the Dordogne and Garonne rivers, and the Gironde Estuary are from several visits to the region by the author in fall 2005.

—Historical facts on Bordeaux, St.-André-de-Cubzac, and Aquitaine are from the author's visits to museums and libraries in the region; accounts in Delie Muller and Jean-Yves Boscher, *Bordeaux: Aspects of Aquitaine* (Bourdeaux: Editions Grand Sud, 2003); and tourist brochures for St.-André-de-Cubzac.

—Facts about Daniel Cousteau's service to James Hazen Hyde are from Madsen, *Cousteau;* Richard Munson, *Cousteau: The Captain and His World* (New York: Paragon House, 1989); and accounts in many magazines about the birth of Jacques Cousteau.

—Cousteau's birth in St.-André-de-Cubzac is confirmed by the author's interviews with an administrator at the village hall and reliable local records. The town house across the street from the church is now a pharmacy, but Cousteau's birthplace is commemorated by a plaque on the building.

—Cousteau's memory of swaying in a hammock on a train is from many interviews and both the Madsen and Munson biographies.

—The description of Eugene Higgins and his high-life activities are from his obituary in the *New York Times,* July 30, 1948.

—The story of Cousteau's first dive into the Vermont lake at the order of Mr. Boetz is recounted in many places, but confirmed in the June 1985 issue of the *Calypso Log,* the publication of the Cousteau Society.

—The clips of Cousteau's earliest attempts at filmmaking were part of a Turner Broadcasting special, *Cousteau: The First Seventy-five Years,* June 1985.

—Information on the several generations of Pathé movie cameras and the history of the Pathé brothers and their company is from several Internet collectors' sites, including http://www.moviecamera.it/pathee.html. Cousteau's camera was undoubtedly the hand-cranked model introduced in 1923. A version with a spring drive was produced in 1926, so by the time he photographed Simone at their first meeting in Paris in 1936 he was probably using that model.

—The accounts of Georges Méliès's discovery of the stop trick and other special effects are from a variety of sources, including Elizabeth Ezra's *Georges Méliès* and *Landmarks of Early Film,* vol. 2, *The Magic of Méliès* (DVD).

—The account of Cousteau being shipped off to boarding school is from several sources, including an interview with him that was the basis for a *Time* magazine story, "Poet of the Depths," which appeared on March 28, 1960. As recited by Cousteau himself many times, the story is presented solely as one of rescue by the discipline and challenges of the Alsatian boarding school, but it is obvious from peripheral facts, including the departure of his brother from the household, the numbing rituals of French education at the time, and the absence of his father, that he was very much a frustrated teenager and not a young filmmaker conducting an experiment when he broke the school windows.

2: Les Mousquemers

—The account of the meeting of Cousteau and Philippe Tailliez, and Tailliez's insistence that Cousteau swim regularly as part of his recuperation, are from Philippe Tailliez, *To Hidden Depths* (London: Kimber, 1954); and from a biographical article by John Christopher Fine that appeared in *Historical Diver* no. 18 (1998).

—The physical details of water and the oceans are taken from *The National Geographic Atlas of the Ocean,* with thanks to its editor, Dr. Sylvia Earle, and the fantastic team of contributors to this essential work.

—Cousteau's comment about his hunger for new developments in skin diving are from Jacques-Yves Cousteau, with Frédéric Dumas (and James Dugan), *The Silent World* (New York: Harper and Row, 1953), 5.

—The presence of precursory gill slits in embryonic chordates was confirmed in an interview with biologist Dr. Tierney Thys.

—John Guy Gilpatric was born in the United States in 1896, set an airplane altitude record of 4,665 feet when he was sixteen years old, flew for the Lafayette Espadrille in World War I, and became a legendary spearfisher who influenced diving pioneer Hans Hass as well as Tailliez, Cousteau, and Dumas. Gilpatric went on to become a journalist and wrote a famous series of books in which the central character is Mr. Glencannon, the ship's engineer of the SS *Inchcliff Castle*. He also wrote the first book on free diving, *The Compleat Goggler*. Gilpatric died in 1950 by his own hand after first ending the life of his terminally ill wife.

—Cousteau's impressions on his first dive are from *The Silent World*, his book published in 1953 chronicling his early years as an underwater explorer.

—Dumas's account of meeting Philippe Tailliez is quoted by Cousteau in *The Silent World*.

—An explanation of atmospheric pressure and breathing can be found in many standard sources, but mine is informed by an interview with diving master Phil Nuytten in Vancouver, British Columbia, November 2005.

3: Breathing Underwater

—Cousteau's exploration of ways to extend his time underwater is from *The Silent World*, 10.

—The survey of pre-Cousteau self-contained underwater breathing apparatuses is from Peter Jackson's article in *Historical Diver* no. 13 (fall 1997), 34–37.

—The description of the Rouquayrol-Denayrouze apparatus is from the Miller and Walter translation of Jules Verne's *20,000 Leagues Under the Sea*, 105–6.

—The account of Cousteau's near-disastrous experiments with an oxygen rebreather is from *The Silent World*, 10–11.

4: *Sixty Feet Down*

—Cousteau's experiments with the surface-feed Fernez diving apparatus and his and Dumas's brushes with death while testing it are from *The Silent World*, 11–12; and Leslie Leaney, "Jacques-Yves Cousteau: The Pioneering Years," *Historical Diver* (fall 1997).

—The insistence of Cousteau's commanders that he continue his diving and underwater photography is from *The Silent World*, 11–12.

—The history of the Williamson brothers and their pioneering work in underwater cinematography is from Thomas Burgess, "The Men Who Made Undersea Films," in *Take Me Under the Sea* (Salem, Oreg.: Ocean Archives, 1994), 163–244, and other sources cited in that essay.

—The purchase of the Kinamo 35 mm camera and Veche's building the underwater housing for it is confirmed in Tailliez, *To Hidden Depths*, 30–32; and Leaney's article. The description of the camera and housing is from scenes included in *Cousteau: The First Seventy-five Years* (June 1985), a Turner Broadcasting special honoring Cousteau.

—The details of shooting *Par dix-huit mètres de fond* (*Sixty Feet Down*) are from Tailliez, *To Hidden Depths*, 30–32.

—The premiere of *Sixty Feet Down* before a gathering of German officers and Vichy French officials in occupied Paris is reported in an article by John Lichfield in the *Independent* (London), June 26, 1999, "20,000 Lies Under the Sea: The Fishy World of Jacques Cousteau."

—The description of the scuttling of the French fleet on November 28, 1942, and the list of ships destroyed is from Cousteau's *The Silent World,* 25–26; and Tailliez's *To Hidden Depths,* 32–33.

5: Scuba

—The epigraph is from an interview with Sara Davidson that appears in a profile in the *New York Times Magazine,* September 10, 1972.

—Biographical information on Émile Gagnan is from Phil Nuytten's article on the invention of the Aqua-Lung in *Historical Diver* (winter 2005).

—The descriptions of the changes in states of matter among liquid, gas, and solid are from Bassam Z. Shakhashiri, "Science Is Fun," http://scifun.chem.wisc.edu/chemweek/Airgases/airgases.html, July 2006.

—The account of Cousteau's first meeting with Émile Gagnan is from an interview with Phil Nuytten and his article "Emil Gagnan and the Aqua-Lung," *Historical Diver* (winter 2005).

—The details of the single-stage regulator are from an interview with engineer and dive historian Phil Nuytten, July 2006; from Nuytten's article in *Historical Diver* (winter 2005); and from Cousteau's *The Silent World,* 12–13. Dive master John Chatterton reviewed my description of the mechanism in the regulator for clarity and accuracy.

—Since Air Liquide had contributed both Gagnan's time and the original design for the natural gas regulator, the corporation owned the patents to the prototype Aqua-Lung.

—The descriptions of Cousteau's Villa Barry in 1941 and 1942 are from Cousteau, *The Silent World,* 11–12; and Tailliez, *To Hidden Depths,* 31–32.

—The account of the arrival of the Aqua-Lung, the first test in the Mediterranean at Bandol, and Cousteau's comparison of a scuba diver and a helmet diver are from Cousteau, *The Silent World,* 1–5.

—The account of Georges Commeinhes's dive off Marseille on July 30, 1943, is from Daniel David, "The Modest Pioneer," *Historical Diving Times,* the newsletter of the Historical Diving Society in Great Britain, summer 1997.

6: Shipwrecks

—Philippe Tailliez's observations on the joy of swimming with the first Aqua-Lung are from his *To Hidden Depths,* 35.

—Cousteau's remembrance of the eight years of goggle diving with Tailliez and Dumas is from *The Silent World,* 5.

—Cousteau's recollection of scavenging for food and the caloric price of scuba diving is from *The Silent World,* 15.

—The account of the dive on *Dalton* and Cousteau's remembrance of Dumas's appetite for loot from the shipwreck is from Cousteau, *The Silent World,* 15–17 and 30–33; and Tailliez, *To Hidden Depths,* 44–45.

—The location of *Dalton* off Planier Island is confirmed on a map in *To Hidden Depths,* 39–40. Tailliez reports that it sank on February 18, 1928; Cousteau reports that it went down on Christmas night, 1928.

—Details of the experiments of Paul Bert and John Scott Haldane are from several sources, including the British Technical Diving Web site Dive Tech: Nitrox and Technical Diver Training, "Decompression Theory: Paul Bert and John Scott Haldane," http://www.dive-tech.co.uk/bert%20and%20haldane.htm. Incredibly, no major book has been written on these pioneers of undersea exploration.

—John Scott Haldane's decompression tables were first published with coauthors A. E. Boycott and G. C. C. Damant in "The Prevention of Compressed Air Illness," *Journal of Hygiene* 8 (1908), 342–443.

—The accounts of the dives to the wrecks of *Tozeur, Ramon Membru, Polyphème,* and *Ferrando,* and Cousteau's remembrances of those dives, are from Cousteau, *The Silent World,* 29–37.

—The account of Dumas's dive to 210 feet is from Tailliez, *To Hidden Depths,* 46–48; and Cousteau, *The Silent World,* 21–23.

7: The Fountain

—The account of the dive into the Fontaine-de-Vaucluse is from Tailliez, *To Hidden Depths,* 73–85; and Cousteau, *The Silent World,* 53–65. The quote from the scientist about the enigma of the fountain is in *To Hidden Depths,* 75.

—Details of the First International Film Festival at Cannes in 1946 are from Peter Burt, ed., *Cannes: Fifty Years of Sun, Sex, and Celluloid* (New York: Miramax, 1997), 2–14.

8: Menfish

—The account of Pierre-Antoine Cousteau's wartime activities and his anti-Semitic writing are from his own books. Cousteau's comment about his brother is from Hugo Frey, *Louis Malle* (Manchester, U.K.: Manchester University Press, 2004), 76; and an article in the (London) *Independent,* "20,000 Lies Under the Sea: The Fishy World of Jacques Cousteau," by John Litchfield, June 26, 1999.

—The improvements in the Cousteau/Gagnan regulator and the change in the housing from Bakelite to metal are documented in Phil Nuytten's indispensable illustrated article "Émile Gagnan and the Aqua-Lung: 1948–1958," *Historical Diver* 13, issue 1, no. 42 (winter 2005).

—The account of the death of Maurice Fargues is from secondary accounts in Munson, *Cousteau: The Captain and His World,* 56; and Madsen, *Cousteau: An Unauthorized Biography,* 58–59. Cousteau, Dumas, and Tailliez make no mention of Fargues's death in their accounts of that period in their books, *The Silent World* and *To Hidden Depths.*

—The account of the Tunisia expedition is from Cousteau, *The Silent World,* 75–85; and Tailliez, *To Hidden Depths,* 85–97.

—The performance of the scouring hoses and the list of artifacts removed from the Roman shipwreck is from Tailliez, *To Hidden Depths,* 95–96; and Cousteau, *The Silent World,* 82–85.

—James Dugan's first article on Cousteau and the Aqua-Lung appeared in *Science Illustrated,* December 1948.

—The history of the arrival of Émile Gagnan in Canada and the emergence of the Aqua-Lung in North American markets is from Nuytten, "Émile Gagnan and the Aqua-Lung: 1948–1958."

—The descriptions of the photographs and text in *Life* magazine are from the November 27, 1950, issue, 119–25.

9: The Abyss

—The description of the headquarters of the Undersea Research Group in Toulon is from Cousteau, *The Silent World,* 49–50.

—Cousteau's enthusiasm for helping Piccard build his deep-diving craft is from Cousteau, *The Silent World,* 95.

—Piccard's comment on the similarity of the ocean depths to the surface of the moon is from his son Jacques Piccard's book, *Seven Miles Down: The Story of the Bathyscaphe "Trieste,"* 6.

—The account of Cousteau breaking his ankle playing tennis with his son Jean-Michel is from an interview with Jean-Michel Cousteau in January 2009.

—The account of the bathyscaphe expedition is from Cousteau, *The Silent World,* 94–101.

10: *Calypso*

—The description of events leading up to the purchase of *Calypso* with help from Loel Guinness is from Cousteau and Sivirine, *Jacques Cousteau's "Calypso,"* 15–17.

—The specifications, history, and details of *BYMS-26* are from Cousteau and Sivirine, *Jacques Cousteau's "Calypso,"* 8–16.

—Cousteau describes his meeting with the admiral in *The Living Sea,* 22.

—Cousteau recounts his conversation with Loel Guinness in *The Living Sea*, 23.

—The descriptions of *Calypso* after modifications in 1950–51 are from Cousteau and Sivirine, *Jacques Cousteau's "Calypso,"* 30–41; and Cousteau, *The Living Sea*, 24–25.

—The account of the sea trials to Corsica is from Cousteau and Sivirine, *Jacques Cousteau's "Calypso,"* 17–18; and Cousteau, *The Living Sea*, 25–26.

11: *Il Faut Aller Voir*

—Cousteau's description of the men who gravitated to him and *Calypso* is quoted in Leslie Leaney, "Jacques-Yves Cousteau: The Pioneering Years," *Historical Diver* (fall 1997).

—The account and details of *Calypso*'s inaugural expedition to the Red Sea is from Cousteau's own account in *The Living Sea*, 26–47.

—Philippe Tailliez would never again be part of the inner circle of Cousteau's life. There is no evidence of animosity between the two men, rather a simple parting of the ways as Tailliez remained in the navy as the commander of the Office of Undersea Technology and Cousteau went on to fame with *Calypso*. In 1954, Tailliez published his own book, *To Hidden Depths*, covering essentially the same material as Cousteau's best seller *The Silent World*. He did not employ a ghostwriter. Cousteau wrote an introductory note: "My dear Tailliez, The memories of the first steps which Dumas, you and I took together under water, and our raptures and our fears, are rooted in the bottom of our hearts. From the first moment your fervor sustained and inspired our team. *To Hidden Depths* will be the living reflection of the spirit which you instilled in us and in the Group." *To Hidden Depths* did not become a best-selling book.

—Cousteau tells the story of the storm over the Malpan Trench in *The Living Sea*, 30–31.

—Cousteau's agreement with James Dugan to ghostwrite *The Silent World* is reported in Madsen, *Cousteau*, 74–76, and confirmed in the author's preface to the book.

—The deal with *Omnibus* on CBS-TV is from Madsen, *Cousteau*, 76, from interviews with Perry Miller reported in the notes for the book.

—Cousteau's dive to search for the amphorae off Grand-Congloue is from Cousteau, *The Living Sea*, 49–50.

—Cousteau describes the salvage operation off Grand-Congloue and the death of Jean-Pierre Servanti in *The Living Sea*, 48–98; and in *Jacques Cousteau's "Calypso,"* 21–25.

—Albert Falco's account of Cousteau inventing the basic design of a research submarine by clapping two saucers together at lunch is quoted in Susan Schiefelbein's introduction to Cousteau and Schiefelbein, *The Human, the Orchid, and the Octopus*, 15.

—Cousteau's proposal to André Laban that they begin work on a submarine that could be launched from *Calypso* and reach depths of 1,000 feet is from an interview with Laban in February 2006.

12: Fame

—Cousteau's association with James Dugan and Dugan's work on *The Silent World* is confirmed in Cousteau's introduction to the book; and in *The Living Sea* (also written with Dugan), 29.

—Rachel Carson's review of *The Silent World* appeared in the *New York Times* on February 8, 1953.

—The account of Cousteau's rescue from financial ruin by D'Arcy Explorations and British Petroleum is from Cousteau, *The Living Sea*, 166–67.

—Louis Malle's arrival as *Calypso*'s chief cameraman and later codirector of *The Silent World* and his participation in the expeditions from 1954 to 1956 is confirmed in Cousteau, *The Living Sea*, 139–76. Malle's career as a director would eventually include: *Ascenseur pour l'échafaud* (*Elevator to the Gallows*), starring Jeanne Moreau as a woman whose lover kills her husband; *Au revoir les enfants* (*Goodbye, Children*) and *Lacombe, Lucien,* stories of betrayal and collaboration with the enemy set in occupied France; *Les amants* (*The Lovers*), in which a woman, played by Jeanne Moreau, abandons her child and husband after one night of sexual abandon with a younger man; and many other tragedies and comedies in France. In 1960 Malle declared that he was "tired of actors, studios, fiction, and Paris," and became a documentarian with a television series on India, after which he moved to the United States. He returned to France to make two more films, one of them a comedy, *Le souffle au coeur* (*Murmur of the Heart*), which includes a scene in which a boy spends a sensual night with his mother. It was roundly condemned in France but hailed by American critic Pauline Kael as a masterpiece and nominated for an Academy Award. The second film after his repatriation to French cinema, *Black Moon,* was a complete flop. Soon after, Malle again moved to the United States, where he directed *Pretty Baby,* about the life of a child in a New Orleans brothel; *Atlantic City,* about a small-time gangster and a casino waitress; and the celebrated *My Dinner with Andre,* a 110-minute conversation in a restaurant between the actors Andre Gregory and Wallace Shawn on the topics of theater, art, and Western civilization.

—The demise of the Persian Gulf pearl diving industry is from Daniel Yergin's great book, *The Prize: The Epic Quest for Oil, Money, and Power* (New York: Free Press, 1991), 293–94.

—The brief history of the development of offshore petroleum exploration and production is from Yergin, *The Prize,* 234–35.

—The encounter with whales off the coast of Africa is from Cousteau, *The Living Sea,* 131–35; and Cousteau and Sivirine, *Jacques Cousteau's "Calypso,"* 29–30.

—The section on the release of *The Silent World* is compiled from the *New York Times,* September 26, 1956; and Cousteau, *The Living Sea,* 139–50.

—The details of Prince Rainier's offer to name Cousteau the director of the Oceanographic Museum of Monaco are from Cousteau, *The Living Sea*, 300–13.

13: Living Underwater

—The account of the design, construction, and testing of *La Souscoupe* is from an interview with André Laban in February 2006; Cousteau, *The Living* Sea, 278–90; and Cousteau and Sivirine, *Jacques Cousteau's "Calypso,"* 51–55.

—André Laban's quote about Cousteau using two saucers to illustrate his design for *La Souscoupe* is from an interview with Laban in February 2006.

—The death of Pierre-Antoine Cousteau was reported in an obituary in the *New York Times* in December 1959. His reconciliation with Jacques Cousteau before he died was related to me in interviews with JYC's son, Jean-Michel, in April 2008, and his grandson, Fabien Cousteau, in February 2005.

—Cousteau's appointment as director of the Oceanographic Museum in Monaco, his planning of the Marine Biotron, and his participation in a conference on nuclear waste are confirmed in Cousteau, *The Living Sea*, 300–313.

—Accounts of the 1959 expedition to photograph the mid-Atlantic rift zone, attendance at the World Oceanographic Congress in New York, and visits to Woods Hole and Washington, D.C., are from Cousteau and Sivirine, *Jacques Cousteau's "Calypso,"* 58–60.

—The article "Poet of the Depths" appeared in *Time,* March 28, 1960.

—The accounts of testing *La Souscoupe* off Puerto Rico, the Cape Verde Islands, and Corsica are from Cousteau, *The Living Sea,* 280–90.

—Cousteau's statement about the importance of human beings actually living in the sea is from his book *World Without Sun,* 6.

—The account of the week Falco and Wesly spent in Conshelf I, and the quotes from them and Cousteau on the experience, are from Cousteau, *The Living Sea,* 314–25.

14: *World Without Sun*

—Cousteau's fantastic prediction of the evolution of *Homo aquaticus* is from contemporary newspaper reports on the World Congress of Underwater Activities in October 1962.

—Cousteau's pronouncement about the importance of the Conshelf II site off Sudan is from Cousteau, *World Without Sun,* 7.

—The account of Conshelf II is from Cousteau and Sivirine, *Jacques Cousteau's "Calypso,"* 73–80.

—Cousteau's fiscal philosophy is inferred from his constant references to raising money in his many books. A conversation with his grandson, Fabien Cousteau, in

February 2005 confirms that Cousteau cared nothing for money itself as long as he had enough to do what he wanted to do.

—The account of Conshelf III is from Cousteau and Sivirine, *Jacques Cousteau's "Calypso,"* 82–85.

—The deal between *National Geographic* and CBS to broadcast *The World of Jacques Cousteau,* edited by David Wolper, is reported in Wolper's autobiography, *Producer: A Memoir* (New York: Scribner, 2003), 112–14.

15: The Undersea World of David Wolper

—The accounts of David Wolper watching Cousteau on television, thinking his undersea adventures would make a great series, his eventual offer to Cousteau, and ABC's Tom Moore agreeing to air twelve episodes are from Wolper's autobiography, *Producer,* 112–17; and Cousteau and Sivirine, *Jacques Cousteau's "Calypso,"* 87–88. Wolper's remark about getting a hold of "that little Frenchie" was passed on to me by Jean-Michel Cousteau in an interview in January 2009.

—David Wolper's story of his breakthrough with *The Race for Space* is from videotaped interviews with him for the Archive of American Television, a division of the Academy of Television Arts and Sciences. Wolper and fellow television producer Grant Tinker are the founding cochairs of the archive. Two hundred of the five hundred interviews with major figures in the history of television are available in full online at http://tvinterviewsarchive.blogspot.com/. Wolper also tells the story in his autobiography, *Producer,* 29–39.

—In a career that continues in 2008, David Wolper has produced more than seven hundred documentaries, feature films, and television series, which have won two Oscars, fifty Emmys, seven Golden Globes, and five Peabodys. His credits include *Funny Bunnies; Hollywood, the Golden Years; The Rafer Johnson Story; The Making of the President, 1960;* the National Geographic Society Specials, 1965–75; *Highlights of the Ice Capades; The Bridge at Remagen; I Love My Wife;* the George Plimpton Specials, 1971–72; *Willie Wonka and the Chocolate Factory; I Will Fight No More Forever; Victory at Entebbe; Roots; This Is Elvis; L.A. Confidential;* and *Celebrate the Century.* His several companies have trained a flood of his protégés whose work has dominated film and television documentary production for fifty years. A complete list of Wolper's films is available at http://www.davidlwolper.com/.

—Cousteau's statement about the truth in poetry is from Wolper, *Producer,* 115.

—Jean-Michel Cousteau's recollection of the fortunate coincidences involved in the deal making for *The Undersea World* is from an interview in April 2008.

—Cousteau's strained but productive relationship with his sons, Philippe and Jean-Michel, is confirmed in interviews with Fabien Cousteau in February 2006 and Jean-Michel in April 2008, as well as in Madsen, *Cousteau,* 144–46.

—Cousteau recounts *Calypso*'s departure from Monaco in February 1967 in *Jacques Cousteau's "Calypso,"* 90–92.

—Philippe Cousteau's presence aboard *Calypso* during the filming of the first episodes of the series and the inventory of camera equipment are confirmed in *The Shark: Splendid Savage of the Sea,* which he coauthored with his father, 1–19.

16: An Honest Witness

—Cousteau's hopes for *The Undersea World* are from reports in the *New York Times* and the *Washington Post,* January 1968.

—The reviews quoted from *Time, Saturday Review,* and *Variety* are from a promotional brochure published by Metromedia, the successor to David Wolper Productions as the producer of *The Undersea World of Jacques Cousteau* after the first four episodes, as cited in Munson, *Cousteau,* 115n.

—Eugenie Clark's critical comments about Cousteau are from the *Washington Post Magazine,* January 11, 1981. Cousteau's statement about his intentions as an honest witness are from an article he wrote for the *New York Times Magazine,* September 10, 1972.

—The chronology of *Calypso*'s voyage from South Africa to the Caribbean with the sea lions is from Cousteau and Sivirine, *Jacques Cousteau's "Calypso,"* 92–119.

—Christopher Palmer's comments about Cousteau's contribution as a pioneer natural history filmmaker are quoted in a *New York Times* article after Cousteau's death in 1997.

—Jean-Michel Cousteau's assessment of his father as a storyteller is from an interview in January 2009.

—Cousteau's comment about his father's death is from an interview in the *Seattle Weekly,* November 1977.

—Cousteau's pronouncement that the oceans are in danger of dying is from an interview published in *Time,* September 28, 1970.

17: *Oasis in Space*

—The details of the foundation of the Cousteau Society are from an interview with Jean-Michel Cousteau in July 2008.

—Biographical information on ARCO chairman Robert Anderson is from Yergin, *The Prize,* 570–71.

—Cousteau's powerful attraction to women and theirs to him were well known among his intimates and associates, including his grandson Fabien Cousteau and *Calypso* crewmen André Laban and Marc Blessington, who shared their observations with me in interviews. Public knowledge of Cousteau's womanizing was limited, but some articles have appeared in European newspapers that treated his indiscretions with the disrespect that is usually the fare of tabloids, though together they serve to confirm it.

—Cousteau and Francine diving together in Houston when Cousteau suffered an earache is confirmed in an interview with Jean-Michel Cousteau in May 2008, and dated by coverage of Involvement Day in the *Houston Chronicle* in September 1977.

18: *Odyssey*

—The details of the sinking of the cargo ship *Cavtat* are from a paper by G. Tiravanti and G. Boari, "Potential Pollution of a Marine Environment by Lead Alkyls: The *Cavtat* Incident," *Environmental Science & Technology* 13 no. 7 (1979), 849–54.

—The description of "Time Bomb at Fifty Fathoms" is from the DVD of the episode released in 2005 by the Cousteau Society and Warner Video.

—The details of the PBY-6A are from Roscoe Creed, *PBY: The Catalina Flying Boat* (Annapolis, Md.: U.S. Naval Institute Press, 1986).

—The account of the crash of *Flying Calypso* and the death of Philippe Cousteau are from several newspaper reports and an obituary in the *New York Times* on July 1, 1979.

—The negative reaction of some of *Calypso*'s crew to Cousteau's remarks to the press after Philippe's funeral is confirmed in an interview with André Laban in February 2006. Laban himself was also offended that Cousteau had filmed the service aboard the Portuguese corvette, and from then on did not consider Cousteau to be a friend.

19: Moving On

—The accounts of life in the Cousteau family after the death of Philippe are from interviews with Fabien Cousteau in February–May 2005 and Jean-Michel Cousteau in April and July 2008.

—Jean-Michel Cousteau's reaction to the death of his brother and subsequent willingness to become his father's second in command are from an interview in July 2008. Quotes from Jean-Michel and Jacques Cousteau about Jean-Michel's assumption of duties at the Cousteau Society are from Madsen, *Cousteau*, 205–7, cited from newspaper reports in 1979–80.

—*Calypso*'s voyage on charter to survey the waters at the mouth of the Orinoco River off Venezuela are from Cousteau and Sivirine, *Jacques Cousteau's "Calypso,"* 166–68.

—The account of the Canadian Film Board grant and subsequent voyage of *Calypso* through the St. Lawrence Waterway is from Cousteau and Sivirine, *Jacques Cousteau's "Calypso,"* 169–79.

—The account of the voyage of *Moulin à Vent* and Cousteau's reaction to its failure are from a United Press International item, November 17, 1983, and a story in *Sail* magazine in February 1984.

20: Captain Outrageous

—*Calypso*'s damage from the St. Lawrence expedition is described in Cousteau and Sivirine, *Jacques Cousteau's "Calypso,"* 181.

—The $5.1 million debt of the Cousteau Society in 1980 was confirmed in an interview with Jean-Michel Cousteau in July 2008.

—The details of the grim meeting in New York, John Denver's suggestion that Cousteau meet Ted Turner, their introduction, and the deal for the Amazon expedition and the rights to the *Odyssey* series are from an interview with Jean-Michel Cousteau, July 2008.

—The details of preparation for the Amazon expedition and repairs to *Calypso* are from Cousteau and Sivirine, *Jacques Cousteau's "Calypso,"* 181–84.

—The details of the Mississippi expedition are from interviews with Jean-Michel Cousteau in July 2008, Fabien Cousteau in February–May 2006, and *Calypso/Alcyone* crew member Marc Blessington in December 2005.

—The facts and details about the arrival of *Alcyone* and *Calypso* in New York in June 1985 are from many newspaper accounts that week.

—President Ronald Reagan's awarding of the Medal of Freedom to Cousteau was reported in the *Washington Post,* April 8, 1985, and June 20, 1985.

—The details of Cousteau's seventy-fifth birthday celebration at Mount Vernon are from a DVD of the WTBS television special *Cousteau: The First Seventy-five Years,* November 1985.

—Ted Turner's remark that Cousteau is the father of the environmental movement is from his autobiography, *Call Me Ted* (New York: Grand Central Publishing, 2008), 210–11.

21: Rediscovering the World

—The story about Ted Turner appearing to fall asleep during the pitch for the *Rediscovery of the World* series was told to me by Fabien Cousteau in February 2006.

—Details of the visit to Haiti are from Munson, *Cousteau: The Captain and His World,* 212–14.

—The details of life aboard *Alcyone* and *Calypso* during *The Rediscovery of the World* series are from conversations with former crewman Marc Blessington in December 2005, Fabien Cousteau between February and May 2006, and Jean-Michel Cousteau in April and July 2008.

—Turner's extension of Cousteau's contract through 1992 and the appointment of Tom Beers as producer were confirmed in an interview with Beers in September 2008.

—Jean-Michel Cousteau's hearing of his mother's death in a phone call while he was in Bangkok is from an interview in July 2008.

22: Chaos

—The accounts of the lunch at which Cousteau broke the news about Francine Triplet to Jean-Michel, and Jean-Michel's subsequent reaction, which led to his quitting work with his father, are from interviews with Jean-Michel Cousteau in April and July 2008, as well as Jacques Cousteau's obituary in the *New York Times,* June 26, 1997.

—Cousteau's musings on his own death and other pronouncements on humanity and the environment are quoted in Madsen, *Cousteau,* 235–39.

—Jean-Michel's observations on his father's last years are from interviews with him in April 2008 and January 2009.

—The account of Cousteau's funeral at Notre Dame is from an article in the *New York Times* by Marlise Simmons, July 1, 1997.

—The account of the dispute over ownership of *Calypso* is from interviews with Fabien Cousteau, February–May 2006, and Jean-Michel Cousteau, April and July 2008. Francine Cousteau's statements about the dispute and its resolution are from the Cousteau Society announcements and an article in the London *Sunday Times,* January 27, 2008.

BIBLIOGRAPHY

Books

Bass, Thomas A. *The Newtonian Casino.* London: Penguin Books, 1990.

Bibb, Porter. *It Ain't As Easy As It Looks: A Biography of Ted Turner.* Boulder, Colo.: Johnson Books, 1997.

Broad, William J. *The Universe Below: Discovering the Secrets of the Deep Sea.* New York: Simon & Schuster, 1997.

Brunner, Bernd. *The Ocean at Home: An Illustrated History of the Aquarium.* New York: Princeton Architectural Press, 2005.

Burgess, Thomas. *Take Me Under the Sea: The Dream Merchants of the Deep.* Salem, Ore.: Ocean Archives, 1994.

Burt, Peter, ed. *Cannes: Fifty Years of Sun, Sex, and Celluloid.* New York: Miramax, 1997.

Chaikin, Andrew. *A Man on the Moon.* New York: Penguin, 1995.

Cousteau, Jacques-Yves, and Philippe Cousteau. *The Shark: Splendid Savage of the Sea.* New York: Doubleday, 1970.

Cousteau, Jacques-Yves, and Philippe Diole. *Octopus and Squid: The Soft Intelligence.* New York: Doubleday, 1973.

Cousteau, Jacques-Yves, with James Dugan. *The Living Sea.* New York: Harper and Row, 1963.

———. *World Without Sun.* New York: Harper and Row, 1964.

Cousteau, Jacques-Yves, with Frédéric Dumas (and James Dugan). *The Silent World.* New York: Harper and Row, 1953, 1981. Reprint edition by the National Geographic Society, 2004.

Cousteau, Jacques-Yves, and Robert Laffont. *The Cousteau Encyclopedia of the Ocean.* Paris and New York: World Publishing (Volumes 1–7), Abrams (Volumes 8–20), 1966–76.

Cousteau, Jacques-Yves, and Susan Schiefelbein. *The Human, the Orchid, and the Octopus: Exploring and Conserving the Natural World.* New York: Bloomsbury USA, 2007.

Cousteau, Jacques-Yves, and Alexis Sivirine. *Jacques Cousteau's "Calypso."* New York: Abrams, 1983.

Cousteau, Jean-Michel. *Mon père, le commandant.* Paris: L'Archipel, 2004.

Cousteau, Pierre-Antoine. *L'Amerique juive.* Paris: Editions de France, 1942.

————. *Après le deluge.* Paris: Editions de France, 1956.

————. *Les lois de l'hospitalité.* Paris: Librairie Française, 1959.

Creed, Roscoe. *PBY: The Catalina Flying Boat.* Annapolis, Md.: U.S. Naval Institute Press, 1986.

Earle, Sylvia. *National Geographic Atlas of the Ocean.* Washington, D.C.: National Geographic Books, 2001.

————. *Sea Change: A Message of the Oceans.* New York: Putnam, 1995.

Ellis, Richard. *The Empty Ocean: Plundering the World's Marine Life.* Washington, D.C.: Islands Press, 2003.

Ezra, Elizabeth. *Georges Méliès.* Manchester, U.K.: Manchester University Press, 2000.

Frey, Hugo. *Louis Malle.* Manchester, U.K.: Manchester University Press, 2004.

Katz, Michael, William P. Marsh, and Gail Gordon Thompson, eds. *Earth's Answer: Exploration of Planetary Culture.* New York: Harper and Row, 1977.

Kelley, Kevin W., ed., for the Association of Space Explorers. *The Home Planet.* New York: Addison-Wesley, 1988.

Kittrell, Ed, Casey Kittrell, and Jim Kittrell, eds. *Down Time: Great Writers on Diving.* 2d ed. Austin, Tex.: Look Away Books, 2002.

Laban, André. *La passion du bleu: Un des pioneers de l'équipe Cousteau raconte.* Aix-en-Province, France: Edisud, 1995.

Madsen, Axel. *Cousteau: An Unauthorized Biography.* New York: Beaufort Books, 1986.

Mander, Jerry. *Four Arguments for the Elimination of Television.* New York: Harper-Collins, 1977.

McIntyre, Joan, ed. *Mind in the Waters: A Book to Celebrate the Consciousness of Whales and Dolphins.* New York: Scribner, 1974.

Muller, Delie, and Jean-Yves Boscher. *Bordeaux: Aspects of Aquitaine.* Bordeaux: Editions Grand Sud, 2003.

Munson, Richard. *Cousteau: The Captain and His World.* New York: Paragon House, 1989.

Piccard, Jacques, and Robert S. Dietz. *Seven Miles Down: The Story of the Bathyscaphe "Trieste."* New York: Putnam, 1961.

Rozwadowski, Helen M. *Fathoming the Ocean: The Discovery and Exploration of the Deep Sea.* Cambridge, Mass.: Belknap Press of Harvard University Press, 2005.

Tailliez, Philippe. *To Hidden Depths.* London: Kimber, 1954.

Thomson, C. Wyville. *The Depths of the Sea.* London: Macmillan, 1874.

Turner, Ted, with Bill Burke. *Call Me Ted.* New York: Grand Central Publishing, 2008.

Verne, Jules. *20,000 Leagues Under the Sea.* Translated by Walter James Miller and Frederick Paul Walter. Annapolis, Md.: U.S. Naval Institute Press, 1993.

Wolper, David L., with David Fisher. *Producer: A Memoir.* New York: Scribner, 2003.

Yergin, Daniel. *The Prize: The Epic Quest for Oil, Money, and Power.* New York: Free Press, 1991.

Periodicals

Calypso Log (Norfolk, Virginia).
Diver (Vancouver, B.C.)
L'Express (Paris)
Historical Diver. Official Publication of the Historical Diving Society, U.S.A.
Historical Diving Times, Newsletter of the Historical Diving Society, Reigate, Surrey, U.K.
National Geographic
New York Times
New York Times Magazine
Saturday Review
Time magazine
Times (London)

Motion Pictures

Landmarks of Early Film. Volume 2: *The Magic of Méliès* (DVD). Image Entertainment, 1999.
Shipwrecks (Épaves). 1943.
The Silent World. 1956.
Sixty Feet Down (Par dix-huit mètres de fond). 1942.
World Without Sun. 1965.

Television

The Undersea World of Jacques Cousteau. ABC, 36 episodes, 1966–75.
Oasis in Space. PBS Canada/United States, 4 episodes, 1976–77.
The Jacques Cousteau Odyssey, PBS United States, 8 episodes, 1977–83.
Amazon and Mississippi. Turner Broadcasting System, 9 episodes, 1978–83.
Jacques Cousteau's Rediscovery of the World, I and *II.* TBS, 36 episodes, 1986–92.

ACKNOWLEDGMENTS

The idea for this book came from the lively mind of Meryl Rosofsky, physician, chef, traveler, bon vivant, and dear friend. I hope I have done justice to her confidence that I could do this job and thank her for her inspiration.

Writing about an internationally celebrated, complex, and beloved man would have been impossible without the generosity and kindness of dozens of people who knew him or in some way counted themselves among his enormous circle of friends and acquaintances. I am most grateful to Fabien Cousteau, who shared my enthusiasm for creating an accurate, engaging portrait of his grandfather, and to Jean-Michel Cousteau, whose forthright recollections informed my understanding of his father. Sandra Squire, Jean-Michel Cousteau's assistant, remained cheerfully helpful even when my many phone calls and requests had certainly become a nuisance. Former members of the crews of *Calypso* and *Alcyone* answered my questions both on and off the record about Cousteau, the inventions he inspired to explore the underwater world, and life aboard ship during forty-five years of filmmaking at sea. Thanks especially to André Laban, Albert Falco, Marc Blessington, and Richard Murphy. Tom Beers, formerly a producer at Turner Broadcasting, provided me with great insights into the rigors and routines of television production as well as some details of the last years of Cousteau's life. Cousteau himself recorded a great deal of his life in words and on film.

My understanding of the development and mechanics of the Aqua-Lung was nurtured by Phil Nuytten, a pioneer diver and renaissance man who never seemed to be annoyed when I asked him the same question two or three times. His explanations of oxygen rebreathers, gas regulators, and the physiology of breathing were crucial to explaining them to readers.

I am grateful to psychologist Dr. Lisa Fortlouis Wood, who helped me understand Cousteau's personality and his pattern of involving countless people in fulfilling his visions and leaving most of them behind as he moved on to his next adventure.

I am not the first writer to attempt to chronicle the life of Jacques Cousteau, and am grateful to all who have gone before me, especially Leslie Leaney and Phil Nuytten. Their finely wrought stories on Cousteau and the Aqua-Lung in *Histor-*

ical Diver magazine were true gifts. I am also indebted to Axel Madsen and Richard Munson, whose biographies of Cousteau guided me through the chronology and events through the 1980s.

Many friends shared this four-year voyage with me, providing advice, comfort, research leads, and recollections of Cousteau. My heartfelt thanks to Barbara Bernstein, Mark Brinster, Gary Burdge, Kelly Cassad, Nicolas Chaubert, Daniel Clem, Gail Cunningham and Sara Wood, Doug Dixon at Pacific Fisherman Shipyard, John Grissim, Brett Hobson, Clyde Hull, Barbara Marrett, Will Nothdurft, Claire Nouvian, Carol Ostrom, Jeff Parkhurst, Bruce Robison, Mark Shelley, Colleen Simpson, Tierney Thys, Ray Troll, and Peter Ward.

My agent, Richard Abate at Endeavor, encouraged me as he always does to think clearly and tell a good story. He led me to an understanding of this book as something other than a recitation of events and always seemed to know what to do when I didn't.

Edward Kastenmeier, my editor, recognized the importance of this work and was patient through some unusual circumstances along the way. I am grateful that he insisted I rise above the ordinary in my perceptions of Cousteau and his significance to the world. I also thank Timothy O'Connell, Katharine Freeman, and the other bookmakers at Pantheon, whose excellent work adds so much to what you now hold in your hands.

I gratefully acknowledge the generous support of the MacDowell Colony and Theodore W. Kheel. Both contributed immensely to my sense of well-being while I worked.

My family, Laara Matsen, Jonas Bendiksen, and Milo Bendiksen, provided me with boundless love and inspiration, ingredients so essential to my life that nothing at all would have happened without them.

BRAD MATSEN
Vashon Island, Washington
February 2009

INDEX

98, 105, 107, 108, 112, 113, 115,
125, 128–29, 136, 144, *149,* 164,
167, 172, 174, 177–89, 203, 211,
213, 217–19, 221, 233, 234,
242–48, 255
*Cousteau: The First Seventy-five
Years,* 234
Cousteau Group, 194
Cousteau Society, xv, 5, 8, 195, 196,
197, 198, 201, 203, 207, 212, 214,
217–18, 221, 223–24, 227, 231,
233, 234, 238, 239, 241, 244, 248,
252, 254, 257
crabs, 216
"Cradle or Coffin?," 201–4
Crete, 118, 207
Cristobal (sea lion), 188–90
Crowther, Bosley, 141, 165–66, 186
crustaceans, 204
Cuba, 240–41
cupric acetate, 103
cyanobacteria, 28

daggers, 76, 79–80
Dalton shipwreck, 64–66, 68, 71,
92, 112
D'Arcy Exploration Co., 128–29, 134
Darwin, Charles, 183
decompression sickness, 66–68, 70,
72, 88–89, 95, 112, 121, 123, 125,
139, 155, 157, 159, 164, 168, 179
decompression stops, 123
Deep Cabin, 163–64, 165
Deguy, Bernard, 241
Delarue, George, 209
Delemotte, Bernard, 219
Delorme, Paul, 53
demand regulators, 54–55
Denayrouze, Auguste, 35
Denver, John, 195, 223–24, *224,*
226, 234
depth gauges, 123
depth soundings, 87
Detroit Diesel Allison factory, 227–28
DeWitt High School, 16
diaphragms, rubber, 54–55, 57

diatoms, 77
Diogenes cylinder, 156–59
DiPerna, Paula, 227
Disneyland, 239
diving:
air supply in, 28–29, 32–38, 41–43
"bends" in, 65–68, 70, 72, 88–89,
95, 112, 121, 123, 125, 139, 155,
157, 159, 164, 168, 179
body temperature in, 31–32, 74, 76
bubbles in, 42, 56–57, 89, 123
buoyancy in, 29, 32, 34, 38, 59, 60,
65, 76
deaths in, 32–33, 125, 219
decompression in, 32–33, 66–68, 70
depth records in, 70–72, 85,
139, 191
duration of, 33
energy loss in, 31–32, 41, 63
equipment for, 25–29, 31–32, 33,
92–93
free swimming (scuba diving) in,
34–35, 41, 42, *42,* 48, 59, 65, 72,
84, 120, 121, 122, 126, 131
hard-hat, 29, 34, 60, 65, 70, 72, 74,
75, 85, 86–87, 93, 132, 155, 158
masks for, 24–26, 29, 31, 42, 59, 75,
76, 159, 206
narcosis in, 71–72, 77–80, 85,
122–23, 139, 206
for oil exploration, 155, 160, 162,
166, 167–68, 176
photographs of, *42, 63, 128*
physical effects of, 38
positions in, 55–57
reconnaissance, 70
research by, 117, 118–19
risks of, 32–33, 34, 38
saturation, 155–59
species hunted in, 25, 29, 31, 33, 37,
41, 47–48, 55, 58, 61, 62, 63, 140
training for, 150–51
for underwater archaeology, 86–90,
94–95, 112, 120, 121–25
visibility in, 37–38
water pressure in, 32–33, 59, 80

flashlights, waterproof, 76, 77, 78
flatfish, 48
Flettner, Anton, 220
Fleuss, Henry, 35–36
floodlights, 101, 135, 138, 139, 179
flying boats, 197
Flying Calypso seaplane, *204,*
    207–12, 239
flying fish, 134
Fontaine-de-Vaucluse, 74–80, 85
Fortune, 225
Forum des Halles, 243
France, Jean-Marie, 227
Free French, 49–50
Free Willy, 254
French Film Institute, 89, 106
French Maritime Museum, xiii, 255
French Ministry of National
    Education, 134
French National Bank, 106
French Oceanographic Expeditions
    (FOE), 111–12, 113, 119, 121,
    122, 134, 256
French Office of Undersea
    Technology, 3, *52,* 124, 128, 135,
    146, 166, 176, 178
French Resistance, 40, 41, 49–50,
    63–64, 82
French Riviera, 31, 33, 58, 91, 121,
    131, 135, 172, 243
Frioul Island, 156
fur seals, 195, 215

Gagnan, Émile, 51–61, 62, 63, 71,
    84–85, 92, 93, 127
Galápagos Islands, 183
Galeazzi decompression chamber, 179
Galliano, Remy, 219
Gary, Roger, 64, 112, 113
gasoline, 54–55, 80, 100, 101, 111, 201,
    204, 226
General Motors, 110
geology, 117, 118–19, 131–33
giant squids, 99
gills, 27, 160
Girault, Yves, 124, 125

Gironde River, 12–13, 253
Glenn, John, 175
goggles, diving, 24–26, 31, 42, 62
Goiran, Henri, 124
gold, 27, 229
golden snakes, 133
Goupil, Pierre, 165
Grace, Princess of Monaco, 177, 178
Grand-Congloue, 121–25, 128, 131,
    133, 135, 145
grappling claws, 99, 100, 101–2
gravimeters, 132
gray whales, 182, 199
grease, 31, 76
Great Barrier Reef, 236–37
Great Britain, 107, 207, 242
Great Lakes, 219
Great Pyramids, 209
Greece, 28, 120, 207
Greenpeace, 195
Gregory, Dick, 195
Grimaldi, Prince Albert, 141, 142
Grimaldi, Prince Louis, 142
Grosvenor, Gilbert, 120–21
groupers, xvi, 26, 33, 121, 140, 152
Guilbert, Pierre, 163, 165
Guilcher, André, 117
Guinness, Loel, 107–10, 112, 256
Gulf of Mexico, 166
Gulf of Oman, 134
Gulf of St. Lawrence, 218, 219
Guralnick, Peter, xvi

Haiti, 240, 251
Haldane, John Scott, 68
halibut, 109
Halifax, Canada, 218–19
Halley, Edmond, 28–29
hammerhead sharks, 180
Hanen, Fernand, 115
harnesses, tank, 57
harpoon cannons, 97, 99, 100, 101–2
harpoons, 25, 29, 33, 97, 199
Heinic, Papa, 46
helium, 66, 86, 95, 97, 164, 165,
    168, 207

recompression treatment, 67–68,
    88–89, 112, 121, 139, 157, 159,
    164, 168, 179
Red Sea, 118–20, 124, 130, 136–37,
    140, 162–67, 178, 179, 181, 242
refrigeration, 51–52
regulators, air, 36, 50, 51, 55–61, 62,
    64, 65, 71, 76, 77, 78–79, 84, 89,
    90, 139
Renoir, Jean, 129
Reynolds, Malvina, 198
Rifkin, Bud, 169, 171–72, 174, 190
Rifkin, Tedde, 172
"rifle clocks," 87, 88
rivers, 226, 227–32, 234–35
Roillet, Jacques, 167
Roman Empire, 7, 12, 70, 86–90,
    121–25, 131
Roosevelt, Franklin Delano, 39–40
Rosaldo, 163
rotor sail system, 220
Rouquayrol, Benoît, 35
Royal Air Force (RAF), 107
Royal Dutch Shell, 132
Royal Navy, 90–91, 110

safety valves, 56
St.-André-de-Cubzac, France, xi,
    xiv–xv, *4,* 12, 13, 14, 15, 114,
    253, *253*
St. Lawrence—Stairway to the Sea, 220
St. Lawrence Waterway, 218–20
St. Pierre Island, 219
salmon, 109, 197
salvage divers, 65, 70, 121, 122
Sanary-sur-Mer, France, 25–26,
    29–30, 31, 34, 36, 38–39, 41, 49,
    58, 72, *74,* 84, 89, 97–98, 102,
    106, 112, 135, 154, 179, 194
Santorini Island, 207
Saout, François, 115, 118
Saturday Review, 186, 204, 215
Scaldis, 100
Scaphandre Autonome, see Aqua-Lung
Schiefelbein, Susan, 215, 227
science fiction, 19, 29, 35, 43, 45, 99,
    103, 107, 115, 160–62

Scientific American, 44
Scientific Illustrated, 92–93, 103
scrap metal, 121, 122
Sea Around Us, The (Carson), 127
Sea Fleas, 176, 179, 191
sea grass, 69
sea lions, 188–90
seals, 127, 195, 215
sea snakes, 133
sea turtles, 134, 180, 184
seawater, 27, 60, 119, 162, 174
seismic equipment, 162, 176
Sephardic Jews, 39–40
Servanti, Jean-Pierre, 124, 125
Seychelles Islands, 136, 137, 138,
    182–83
Shackleton, Ernest, 175
shark cages, 133, 137, 140, 188
sharks, 88, 103, 127, 133, 137–38,
    140, 141, 165, 172, 179–81, 186,
    188, 216
"Sharks," 179–81
shipwrecks, 62–72, 80–81, 85, 86–90,
    94–95, 102–3, 112, 120, 121–25,
    127, 130, 131, 140, 207, 215,
    219, 220
Shoppelry, Richard, 185
shrimp, 158, 192
side-scan sonar, 136
Sierra Club, 195
Silent World, The (Cousteau), 85,
    127–28, 133, 136
Silent World, The (film), xvi, 128, 136,
    138–41, 143, 144, 150, 155, 162,
    165, 178, 179, 186–87, 250
Six-Day War (1967), 181–82
skin bladders, 28–29
slaughterhouses, 52
sledgehammers, 132–33
sleds, towed, 86, 87–88, 147–48
Smith, Allyn, 103
snorkels, 25–26, 29, 33, 59
"Snowstorm in the Jungle," 235
Société Zix, 21
solar power, 242–43
Solt, Andrew, 196
sonar, 114, 136, 146, 151

A NOTE ABOUT THE AUTHOR

Brad Matsen has been writing about the sea and its inhabitants for forty years. He is the author of *Descent: The Heroic Discovery of the Abyss,* which was a finalist for the Los Angeles Times Book Prize in 2006; *Titanic's Last Secrets; Planet Ocean: A Story of Life, the Sea, and Dancing to the Fossil Record;* the award-winning *Incredible Ocean Adventure* series for children; and many other books. He was a creative producer for the *Shape of Life,* an eight-hour National Geographic television series on evolutionary biology, and has written on marine science and the environment for *Mother Jones, Audubon, Natural History,* and many other magazines. He lives on Vashon Island in Puget Sound.

A NOTE ABOUT THE TYPE

This book was set in a version of the well-known Monotype face Bembo. This letter was cut for the celebrated Venetian printer Aldus Manutius by Francesco Griffo, and first used in Pietro Cardinal Bembo's *De Aetna* of 1495.

The companion italic is an adaptation of the chancery script type designed by the calligrapher and printer Lodovico degli Arrighi.

Composed by North Market Street Graphics
Lancaster, Pennsylvania

Printed by Berryville Graphics,
Berryville, Virginia

Designed by M. Kristen Bearse